Learn, Practice, Succeed

Eureka Math®
Grade 7
Module 2

Published by Great Minds®.

Copyright © 2019 Great Minds®.

Printed in the U.S.A.

This book may be purchased from the publisher at eureka-math.org.

10 9 8 7 6 5 4 3 2

ISBN 978-1-64054-973-9

G7-M2-LPS-05.2019

Students, families, and educators:

Thank you for being part of the *Eureka Math®* community, where we celebrate the joy, wonder, and thrill of mathematics.

In *Eureka Math* classrooms, learning is activated through rich experiences and dialogue. That new knowledge is best retained when it is reinforced with intentional practice. The *Learn, Practice, Succeed* book puts in students' hands the problem sets and fluency exercises they need to express and consolidate their classroom learning and master grade-level mathematics. Once students learn and practice, they know they can succeed.

What is in the Learn, Practice, Succeed *book?*

Fluency Practice: Our printed fluency activities utilize the format we call a Sprint. Instead of rote recall, Sprints use patterns across a sequence of problems to engage students in reasoning and to reinforce number sense while building speed and accuracy. Sprints are inherently differentiated, with problems building from simple to complex. The tempo of the Sprint provides a low-stakes adrenaline boost that increases memory and automaticity.

Classwork: A carefully sequenced set of examples, exercises, and reflection questions support students' in-class experiences and dialogue. Having classwork preprinted makes efficient use of class time and provides a written record that students can refer to later.

Exit Tickets: Students show teachers what they know through their work on the daily Exit Ticket. This check for understanding provides teachers with valuable real-time evidence of the efficacy of that day's instruction, giving critical insight into where to focus next.

Homework Helpers and Problem Sets: The daily Problem Set gives students additional and varied practice and can be used as differentiated practice or homework. A set of worked examples, Homework Helpers, support students' work on the Problem Set by illustrating the modeling and reasoning the curriculum uses to build understanding of the concepts the lesson addresses.

Homework Helpers and Problem Sets from prior grades or modules can be leveraged to build foundational skills. When coupled with *Affirm®*, *Eureka Math*'s digital assessment system, these Problem Sets enable educators to give targeted practice and to assess student progress. Alignment with the mathematical models and language used across *Eureka Math* ensures that students notice the connections and relevance to their daily instruction, whether they are working on foundational skills or getting extra practice on the current topic.

Where can I learn more about Eureka Math *resources?*

The Great Minds® team is committed to supporting students, families, and educators with an ever-growing library of resources, available at eureka-math.org. The website also offers inspiring stories of success in the *Eureka Math* community. Share your insights and accomplishments with fellow users by becoming a *Eureka Math* Champion.

Best wishes for a year filled with "aha" moments!

Jill Diniz

Jill Diniz
Chief Academic Officer, Mathematics
Great Minds

Contents

Module 2: Rational Numbers

Exercise 1: Positive and Negative Numbers Review

With your partner, use the graphic organizer below to record what you know about positive and negative numbers. Add or remove statements during the whole-class discussion.

Negative Numbers   **Positive Numbers**

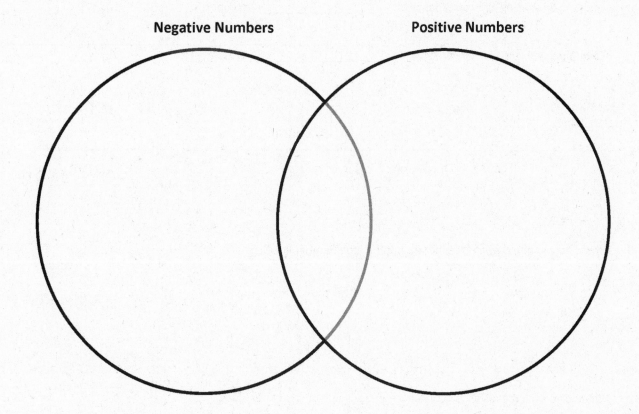

Example 2: Counting Up and Counting Down on the Number Line

Use the number line below to practice counting up and counting down.

- *Counting up* starting at 0 corresponds to _____ numbers.

- *Counting down* starting at 0 corresponds to _____ numbers.

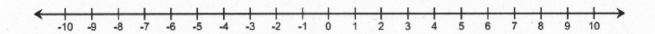

a. Where do you begin when locating a number on the number line?

b. What do you call the distance between a number and 0 on a number line?

c. What is the relationship between 7 and −7?

EUREKA
MATH

Example 3: Using the Integer Game and the Number Line

What is the sum of the card values shown? Use the counting on method on the provided number line to justify your answer.

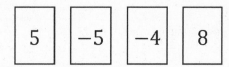

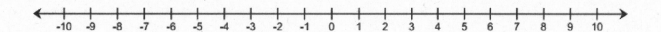

a. What is the final position on the number line? _____

b. What card or combination of cards would you need to get back to 0? _____

Exercise 2: The Additive Inverse

Use the number line to answer each of the following questions.

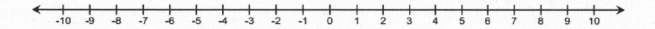

a. How far is 7 from 0 and in which direction? _____

b. What is the opposite of 7? _____

c. How far is −7 from 0 and in which direction? _____

d. Thinking back to our previous work, explain how you would use the counting on method to represent the following: While playing the Integer Game, the first card selected is 7, and the second card selected is −7.

e. What does this tell us about the sum of 7 and its opposite, −7?

f. Look at the curved arrows you drew for 7 and −7. What relationship exists between these two arrows that would support your claim about the sum of 7 and −7?

g. Do you think this will hold true for the sum of any number and its opposite? Why?

Property: For every number a, there is a number $-a$ so that $a + (-a) = 0$ and $(-a) + a = 0$.

The *additive inverse of a number* is a number such that the sum of the two numbers is 0. The opposite of a number satisfies this definition: For example, the opposite of 3 is −3, and $3 + (-3) = 0$. Hence −3 is the additive inverse of 3.

The property above is usually called the existence of additive inverses.

Exercise 3: Playing the Integer Game

Play the Integer Game with your group. Use a number line to practice counting on.

Lesson 1:   Opposite Quantities Combine to Make Zero

EUREKA MATH

Lesson Summary

- Add a positive number to a number by counting up from that number, and add a negative number to a number by counting down from that number.
- An integer plus its opposite sum to zero.
- The opposite of a number is called the additive inverse because the two numbers' sum is zero.

Name _____   Date _____

1. Your hand starts with the 7 card. Find three different pairs that would complete your hand and result in a value of zero.

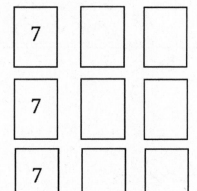

2. Write an equation to model the sum of the situation below.

   A hydrogen atom has a zero charge because it has one negatively charged electron and one positively charged proton.

3. Write an equation for each diagram below. How are these equations alike? How are they different? What is it about the diagrams that lead to these similarities and differences?

Diagram A:

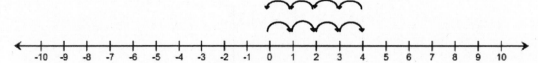

Diagram B:

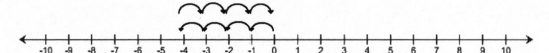

Lesson 1: Opposite Quantities Combine to Make Zero

EUREKA MATH®

Positions on the Number Line

1. Refer to the integer game when answering the following questions.

    a. When playing the Integer Game, what two cards could have a score of −14?

    There are many possible answers. Some of the pairs that make −14 include −10 and −4, −12 and −2, or −15 and 1.

    > I need to start at zero and find two moves so that the second move ends on −14. I can begin my moves with many different numbers.

    b. If the two cards played in a round are the same distance from zero but are on opposite sides of zero, what is the score for the round?

    The two given cards would be opposites, and the score for the round would be zero.

    > Positive cards move to the right of zero, and negative cards move to the left. So if I start at zero and move 5 to the right and then 5 to the left, I will end up back at zero.

2. Hector was given $20 as a gift. He spent $12 at the store and then planned to spend $14 more on a second item. How much more would he need in order to buy the second item? Be sure to show your work using addition of integers.

    Hector would need $6 more.

    $$20 + (-12) + (-14) + 6 = 0$$

    > Hector doesn't have enough money. To have enough money, he needs to end on 0 on the number line. If Hector adds $20 + (-12) + (-14)$, he ends on −6.

    > The money he is given can be positive, and the amount he spends will be negative.

3. Use the 8 card and its additive inverse to write a real-world story problem about their sum.

> An additive inverse is the same distance from zero, but on the opposite side of zero on the number line.

The temperature in the morning was −8°F.

If the temperatures rises 8 degrees, what is the new temperature?

Answer: $(-8) + 8 = 0; 0°F$

> Real-world problems with integers could include money, temperatures, elevations, or even sports.

4. Write an addition number sentence that corresponds to the arrows below.

$$8 + (-3) + (-5) = 0$$

> I start from 0 and can see arrows moving to the right and then to the left. An arrow moving to the right shows a positive addend, and an arrow moving to the left shows a negative addend.

Lesson 1:   Opposite Quantities Combine to Make Zero

EUREKA MATH

For Problems 1 and 2, refer to the Integer Game.

1. You have two cards with a sum of (-12) in your hand.

    a. What two cards could you have?

    b. You add two more cards to your hand, but the total sum of the cards remains the same, (-12). Give some different examples of two cards you could choose.

2. Choose one card value and its additive inverse. Choose from the list below to write a real-world story problem that would model their sum.

    a. Elevation: above and below sea level

    b. Money: credits and debits, deposits and withdrawals

    c. Temperature: above and below 0 degrees

    d. Football: loss and gain of yards

3. On the number line below, the numbers h and k are the same distance from 0. Write an equation to express the value of $h + k$. Explain.

    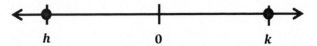

4. During a football game, Kevin gained five yards on the first play. Then he lost seven yards on the second play. How many yards does Kevin need on the next play to get the team back to where they were when they started? Show your work.

5. Write an addition number sentence that corresponds to the arrows below.

Exercise 1: Real-World Introduction to Integer Addition

Answer the questions below.

a. Suppose you received $10 from your grandmother for your birthday. You spent $4 on snacks. Using addition, how would you write an equation to represent this situation?

b. How would you model your equation on a number line to show your answer?

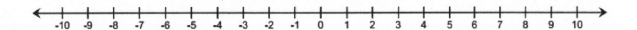

Example 1: Modeling Addition on the Number Line

Complete the steps to find the sum of $-2 + 3$ by filling in the blanks. Model the equation using straight arrows called *vectors* on the number line below.

a. Place the tail of the arrow on _____.

b. Draw the arrow 2 units to the left of 0, and stop at _____. The direction of the arrow is to the _____ since you are counting down from 0.

c. Start the next arrow at the end of the first arrow, or at _____.

d. Draw the second arrow _____ units to the right since you are counting up from -2.

e. Stop at _____.

f.   Circle the number at which the second arrow ends to indicate the ending value.

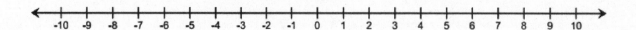

g.   Repeat the process from parts (a)–(f) for the expression 3 + (−2).

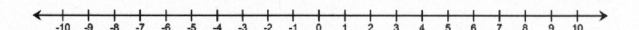

h.   What can you say about the sum of −2 + 3 and 3 + (−2)? Does order matter when adding numbers? Why or why not?

Example 2: Expressing Absolute Value as the Length of an Arrow on the Real Number Line

a.   How does absolute value determine the arrow length for −2? Use the number line provided to support your answer.

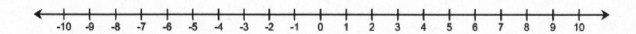

EUREKA
MATH

b. How does the absolute value determine the arrow length for 3? Use the number line provided to support your answer.

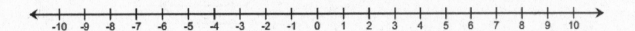

c. Describe how the absolute value helps you represent −10 on a number line.

Exercise 2

Create a number line model to represent each of the expressions below.

a. −6 + 4

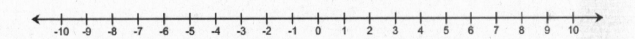

b. 3 + (−8)

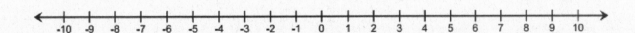

Example 3: Finding Sums on a Real Number Line Model

Find the sum of the integers represented in the diagram below.

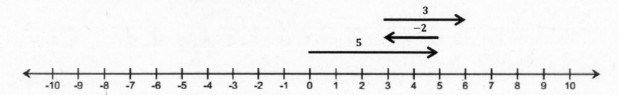

a.   Write an equation to express the sum.

b.   What three cards are represented in this model? How did you know?

c.   In what ways does this model differ from the ones we used in Lesson 1?

d.   Can you make a connection between the sum of 6 and where the third arrow ends on the number line?

e.   Would the sum change if we changed the order in which we add the numbers, for example, $(-2) + 3 + 5$?

f.   Would the diagram change? If so, how?

Exercise 3

Play the Integer Game with your group. Use a number line to practice counting on.

Lesson Summary

- On a number line, arrows are used to represent integers; they show length and direction.
- The length of an arrow on the number line is the absolute value of the integer.
- Adding several arrows is the same as combining integers in the Integer Game.
- The sum of several arrows is the final position of the last arrow.

Name _____ Date _____

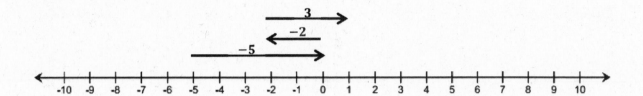

Jessica made the addition model below of the expression $(-5) + (-2) + 3$.

   a. Do the arrows correctly represent the numbers that Jessica is using in her expression?

   b. Jessica used the number line diagram above to conclude that the sum of the three numbers is 1. Is she correct?

   c. If she is incorrect, find the sum, and draw the correct model.

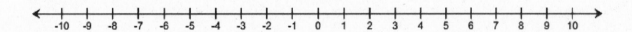

   d. Write a real-world situation that would represent the sum.

Adding Integers on a Number Line

1. When playing the Integer Game, Sally drew three cards, 3, −12, and 8. Then Sally's partner gave Sally a 5 from his hand.

   a. What is Sally's total? Model the answer on the number line and using an equation.

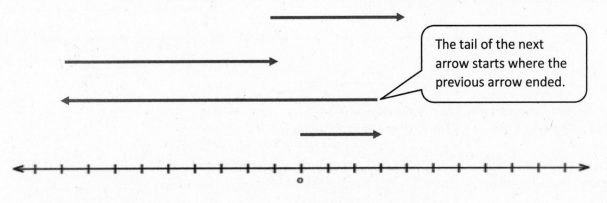

> I use arrows to represent each number. Negative numbers will face left, and positive numbers will face right.

> The tail of the next arrow starts where the previous arrow ended.

$$3 + (-12) + 8 + 5 = 4$$

> I also need to show my work using an equation. I can show all of the numbers being added together. The answer is where the last arrow ends.

   b. What card(s) would you need to get your score back to zero? Explain.

A −4 card would bring the score back to 0. The number −4 is the additive inverse of 4. It is the same distance from 0 on the number line but in the opposite direction. So when I add 4 and −4, the answer would be 0.

> I could also choose more than one card, but the sum must be −4.

2. Write a story problem and an equation that would model the sum of the numbers represented by the arrows in the number diagram below.

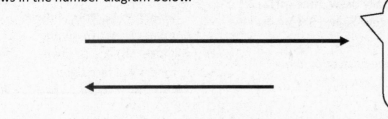

> The bottom arrow is 8 units and faces left. So it shows −8. The second arrow is 11 units and faces to the right. So that arrow represents 11.

In the morning, I lost $8. Later in the day, I got paid $11. How much money do I have at the end of the day?

$$-8 + 11 = 3$$

I had $3 at the end of the day.

> An equation requires that I also state the answer when the sum is calculated.

3. Mark an integer between −2 and 4 on a number line, and label it point K. Then, locate and label each of the following points by finding the sums.

> My answer will depend on what I pick for K. For this example, I pick −1 for K.

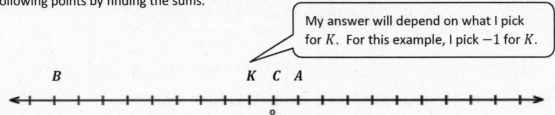

a. Point A: $K + 2$

   Point A: $-1 + 2 = 1$

> I can use K to help me when adding the other numbers. I will always start at K to determine where each point should be located on the number line.

b. Point B: $K + (-8)$

   Point B: $-1 + (-8) = -9$

c. Point C: $(-4) + 5 + K$

   Point C: $(-4) + 5 + (-1) = 0$

Represent Problems 1–3 using both a number line diagram and an equation.

1. David and Victoria are playing the Integer Card Game. David drew three cards, −6, 12, and −4. What is the sum of the cards in his hand? Model your answer on the number line below.

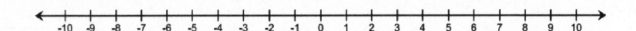

2. In the Integer Card Game, you drew the cards, 2, 8, and −11. Your partner gave you a 7 from his hand.

   a. What is your total? Model your answer on the number line below.

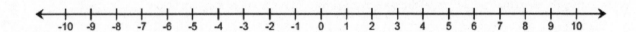

   b. What card(s) would you need to get your score back to zero? Explain. Use and explain the term *additive inverse* in your answer.

3. If a football player gains 40 yards on a play, but on the next play, he loses 10 yards, what would his total yards be for the game if he ran for another 60 yards? What did you count by to label the units on your number line?

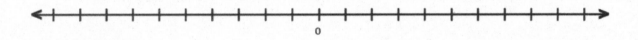

4. Find the sums.

   a. $-2 + 9$

   b. $-8 + -8$

   c. $-4 + (-6) + 10$

   d. $5 + 7 + (-11)$

EUREKA
MATH®

5. Mark an integer between 1 and 5 on a number line, and label it point Z. Then, locate and label each of the following points by finding the sums.

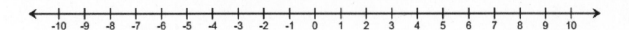

a. Point A: $Z + 5$
b. Point B: $Z + (-3)$
c. Point C: $(-4) + (-2) + Z$
d. Point D: $-3 + Z + 1$

6. Write a story problem that would model the sum of the arrows in the number diagram below.

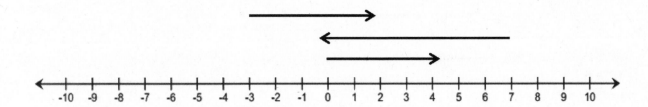

7. Do the arrows correctly represent the equation $4 + (-7) + 5 = 2$? If not, draw a correct model below.

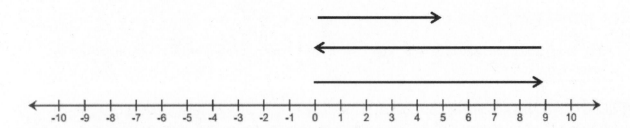

EUREKA MATH

Exercise 1: Addition Using the Integer Game

Play the Integer Game with your group without using a number line.

Example 1: Counting On to Express the Sum as Absolute Value on a Number Line

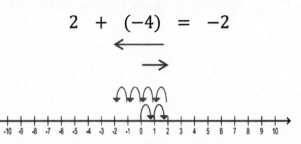

Model of Counting Up

$$2 + 4 = 6$$

Model of Counting Down

$$2 + (-4) = -2$$

Counting up −4 is the same as *the opposite of counting up* 4 and also means counting down 4.

a. For each example above, what is the distance between 2 and the sum?

b. Does the sum lie to the right or left of 2 on a horizontal number line? Above or below on a vertical number line?

c. Given the expression 54 + 81, determine, without finding the sum, the distance between 54 and the sum. Explain.

d. Is the sum to the right or left of 54 on the horizontal number line? Above or below on a vertical number line?

e. Given the expression $14 + (-3)$, determine, without finding the sum, the distance between 14 and the sum. Explain.

f. Is the sum to the right or left of 14 on the number line? Above or below on a vertical number line?

Exercise 2

Work with a partner to create a horizontal number line model to represent each of the following expressions. What is the sum?

a. $-5 + 3$

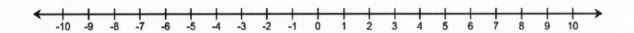

EUREKA MATH

b. −6 + (−2)

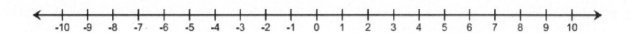

c. 7 + (−8)

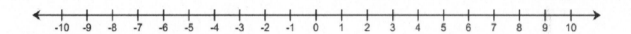

Exercise 3: Writing an Equation Using Verbal Descriptions

Write an equation, and using the number line, create an *arrow diagram* given the following information:

The sum of 6 and a number is 15 units to the left of 6 on the number line.

Equation:

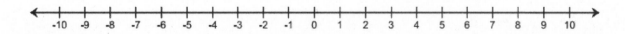

Lesson Summary

- Adding an integer to a number can be represented on a number line as counting up when the integer is positive (just like whole numbers) and counting down when the integer is negative.

- Arrows can be used to represent the sum of two integers on a number line.

EUREKA
MATH®

Name _____   Date _____

1. Refer to the diagram to the right.

    a. Write an equation for the diagram to the right. _____

    b. Find the sum. _____

    c. Describe the sum in terms of the distance from the first addend. Explain.

    d. What integers do the arrows represent? _____

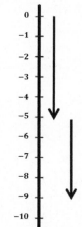

2. Jenna and Jay are playing the Integer Game. Below are the two cards they selected.

    a. How do the models for these two addition problems differ on a number line? How are they the same?

    Jenna's Hand   Jay's Hand

    b. If the order of the cards changed, how do the models for these two addition problems differ on a number line?
        How are they the same?

    Jenna's Hand   Jay's Hand

1. Refer to the diagram to the right.

    a. What integers do the arrows represent?

        6 *and* −14

    b. Write an equation for the diagram to the right.

        6 + (−14) = −8

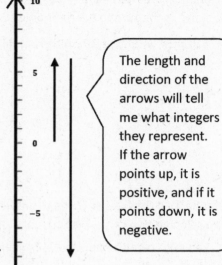

The second arrow is pointing to the sum. So I can write an equation showing the sum of the two integers being equal to the sum.

The length and direction of the arrows will tell me what integers they represent. If the arrow points up, it is positive, and if it points down, it is negative.

    c. Describe the sum in terms of the distance from the first addend. Explain.

        ***The sum is* 14 *units below* 6 *because* | −14 | = 14. *I counted down from* 6 *fourteen units and stopped at* −8.**

    d. Describe the arrows you would use on a vertical number line in order to solve −3 + −9.

        ***The first arrow would start at* 0 *and be three units long, pointing downward because the addend is negative. The second arrow would start at* −3 *and be nine units long, also pointing downward. The second arrow would end at* −12.**

The absolute value of the numbers will give me the length of the arrow, and the sign will tell me what direction the arrow should face.

EUREKA MATH®

2. Given the expression $84 + (-29)$, can you determine, without finding the sum, the distance between 84 and the sum? Is the sum to the right or left of 84 on the number line?

The distance would be 29 units from 84. The sum is to the left of 84 on the number line.

> If I draw a sketch of the sum, I start at 0 and move 84 units to the right. I would then have to move 29 units to the left, which means the sum will be 29 units to the left of 84.

3. Refer back to the Integer Game to answer this question. Juno selected two cards. The sum of her cards is 16.

> If needed, I can draw a number line to see what would happen if both of Juno's cards are negative.

   a. Can both cards be negative? Explain why or why not.

   No. In order for the sum to be 16, at least one of the addends would have to be positive. If both cards are negative, then Juno would count twice going to the left/down, which would result in a negative sum.

   b. Can one of the cards be positive and the other be negative? Explain why or why not.

   Yes. She could have −4 and 20 or −2 and 18. The card with the greatest absolute value would have to be positive.

> I can create a number line to determine if this is possible. This visual will also help me see that the longer arrow (larger absolute value) must be positive to get a positive sum.

4. Determine the afternoon temperatures for each day. Write an equation that represents each situation.

   a. The morning temperature was 8°F and then fell 11 degrees in the afternoon.

$$8 + (-11) = -3$$

   The afternoon temperature will be −3°F.

> I can show the temperature falling as adding a negative because it would show a move down on a vertical number line.

   b. The morning temperature was −5°F and then rose 9 degrees in the afternoon.

$$-5 + 9 = 4$$

   The afternoon temperature will be 4°F.

> I can show the temperature rising as adding a positive because it would show a move up on a vertical number line.

EUREKA MATH

1. Below is a table showing the change in temperature from morning to afternoon for one week.

    a. Use the vertical number line to help you complete the table. As an example, the first row is completed for you.

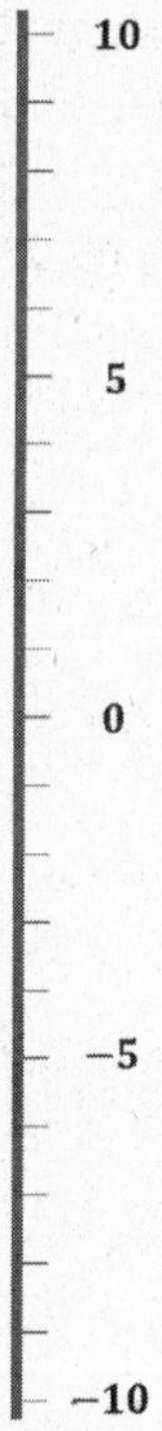

Change in Temperatures from Morning to Afternoon

Morning Temperature	Change	Afternoon Temperature	Equation
1°C	Rise of 3°C	4°C	$1 + 3 = 4$
2°C	Rise of 8°C		
−2°C	Fall of 6°C		
−4°C	Rise of 7°C		
6°C	Fall of 9°C		
−5°C	Fall of 5°C		
7°C	Fall of 7°C		

(Number line ranges from 10 at top to −10 at bottom, marked at 10, 5, 0, −5, −10.)

    b. Do you agree or disagree with the following statement: "A rise of −7°C" means "a fall of 7°C"? Explain. (Note: No one would ever say, "A rise of −7 degrees"; however, mathematically speaking, it is an equivalent phrase.)

For Problems 2–3, refer to the Integer Game.

2. Terry selected two cards. The sum of her cards is −10.

    a. Can both cards be positive? Explain why or why not.

    b. Can one of the cards be positive and the other be negative? Explain why or why not.

    c. Can both cards be negative? Explain why or why not.

3. When playing the Integer Game, the first two cards you selected were −8 and −10.

    a. What is the value of your hand? Write an equation to justify your answer.

    b. For part (a), what is the distance of the sum from −8? Does the sum lie to the right or left of −8 on the number line?

    c. If you discarded the −10 and then selected a 10, what would be the value of your hand? Write an equation to justify your answer.

4. Given the expression 67 + (−35), can you determine, without finding the sum, the distance between 67 and the sum? Is the sum to the right or left of 67 on the number line?

5. Use the information given below to write an equation. Then create an *arrow diagram* of this equation on the number line provided below.

    The sum of −4 and a number is 12 units to the right of −4 on a number line.

Lesson 3:   Understanding Addition of Integers

EUREKA
MATH

Example 1: Rule for Adding Integers with Same Signs

a. Represent the sum of $3 + 5$ using arrows on the number line.

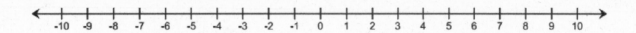

i. How long is the arrow that represents 3?

ii. What direction does it point?

iii. How long is the arrow that represents 5?

iv. What direction does it point?

v. What is the sum?

vi. If you were to represent the sum using an arrow, how long would the arrow be, and what direction would it point?

vii.   What is the relationship between the arrow representing the number on the number line and the absolute value of the number?

viii. Do you think that adding two positive numbers will always give you a greater positive number? Why?

b.   Represent the sum of $-3 + (-5)$ using arrows that represent -3 and -5 on the number line.

i.   How long is the arrow that represents -3?

ii.   What direction does it point?

iii.   How long is the arrow that represents -5?

iv.   What direction does it point?

v.   What is the sum?

vi. If you were to represent the sum using an arrow, how long would the arrow be, and what direction would it point?

vii. Do you think that adding two negative numbers will always give you a smaller negative number? Why?

c. What do both examples have in common?

RULE : *Add rational numbers with the same sign by adding the absolute values and using the common sign.*

Exercise 2

a. Decide whether the sum will be positive or negative without actually calculating the sum.

i. $-4 + (-2)$ _____

ii. $5 + 9$ _____

iii. $-6 + (-3)$ _____

iv. $-1 + (-11)$ _____

v. $3 + 5 + 7$ _____

vi. $-20 + (-15)$ _____

b.   Find the sum.

   i. $15 + 7$

   ii. $-4 + (-16)$

   iii. $-18 + (-64)$

   iv. $-205 + (-123)$

Example 2: Rule for Adding Opposite Signs

a.   Represent $5 + (-3)$ using arrows on the number line.

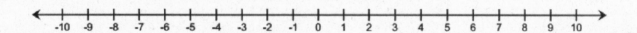

   i.   How long is the arrow that represents 5?

   ii.   What direction does it point?

   iii. How long is the arrow that represents -3?

   iv.   What direction does it point?

EUREKA
MATH®

v.   Which arrow is longer?

vi. What is the sum? If you were to represent the sum using an arrow, how long would the arrow be, and what direction would it point?

b.   Represent the $4 + (-7)$ using arrows on the number line.

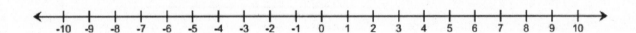

i.   In the two examples above, what is the relationship between the length of the arrow representing the sum and the lengths of the arrows representing the two addends?

ii. What is the relationship between the direction of the arrow representing the sum and the direction of the arrows representing the two addends?

iii. Write a rule that will give the length and direction of the arrow representing the sum of two values that have opposite signs.

RULE: *Add rational numbers with opposite signs by subtracting the absolute values and using the sign of the integer with the greater absolute value.*

Exercise 3

a. Circle the integer with the greater absolute value. Decide whether the sum will be positive or negative without actually calculating the sum.

     i. $-1 + 2$ _____

     ii. $5 + (-9)$ _____

     iii. $-6 + 3$ _____

     iv. $-11 + 1$ _____

b. Find the sum.

     i. $-10 + 7$

     ii. $8 + (-16)$

     iii. $-12 + (65)$

     iv. $105 + (-126)$

EUREKA MATH®

Example 3: Applying Integer Addition Rules to Rational Numbers

Find the sum of $6 + \left(-2\frac{1}{4}\right)$. The addition of rational numbers follows the same rules of addition for integers.

    a.   Find the absolute values of the numbers.

    b.   Subtract the absolute values.

    c.   The answer will take the sign of the number that has the greater absolute value.

Exercise 4

Solve the following problems. Show your work.

    a.   Find the sum of $-18 + 7$.

    b.   If the temperature outside was 73 degrees at 5:00 p.m., but it fell 19 degrees by 10:00 p.m., what is the temperature at 10:00 p.m.? Write an equation and solve.

    c.   Write an addition sentence, and find the sum using the diagram below.

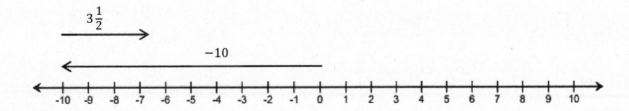

Lesson Summary

- Add integers with the same sign by adding the absolute values and using the common sign.
- Steps to adding integers with opposite signs:
    1. Find the absolute values of the integers.
    2. Subtract the absolute values.
    3. The answer will take the sign of the integer that has the greater absolute value.
- To add rational numbers, follow the same rules used to add integers.

Lesson 4:   Efficiently Adding Integers and Other
Rational Numbers

EUREKA
MATH®

Name _____   Date _____

1. Write an addition problem that has a sum of $-4\frac{3}{5}$ and

    a. The two addends have the same sign.

    b. The two addends have different signs.

2. In the Integer Game, what card would you need to draw to get a score of 0 if you have a -16, -35, and 18 in your hand?

1. Use the diagram below to complete each part.

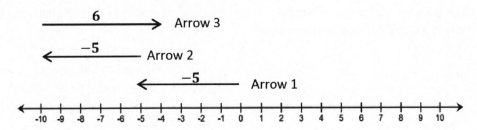

a. How long is each arrow? What direction does each arrow point?

Arrow	Length	Direction
1	5	*left*
2	5	*left*
3	6	*right*

b. Label each arrow with the number the arrow represents.

> I can use the length and direction of each arrow to help me determine what number it represents. If it is facing left, it represents a negative number. If it is facing right, it represents a positive number.

c. Write an equation that represents the sum of the numbers. Find the sum.

$$(-5) + (-5) + 6 = -4$$

> These three arrows represent a sum. The third arrow ends on the answer, or the sum, of all three numbers being represented by the three arrows.

2. Which of these story problems describes the sum $24 + (-17)$? Check all that apply.

___X___ Morgan planted 24 tomato plants at the beginning of spring. She sold 17 of the plants at the farmer's market. How many plants does she have now?

> Selling some of the plants would be represented as a negative number when trying to figure out how many plants Morgan has left.

___X___ Morgan started with 24 tomato plants. Then her mother took 8 of the plants, and her aunt took 9 of the plants. How many plants does Morgan have now?

> If I combine the amounts taken by each family member, I get $(-8) + (-9) = -17$. So this would also show -17 plants being added to the total.

_____ Morgan owes her mother 24 tomato plants but only has 17 to give her. How many tomato plants short of the total needed is Morgan?

> The amount of plants Morgan owes her mom would be represented by a negative number, while the 17 plants she has would be represented by a positive number. Therefore, the signs of the two addends are opposites of the addends in the given expression.

3. Ezekiel is playing the Integer Game. He has the cards -7 and 4.

   a. What card would Ezekiel need to draw next to win with a score of zero?

   > I need to calculate the sum of his two cards.

   $$-7 + 4 = -3$$
   $$-3 + 3 = 0$$

   > If I add a number and its opposite, I will get zero. So I know that the next card drawn has to be the opposite of -3.

   Ezekiel would need to draw a 3.

Lesson 4:   Efficiently Adding Integers and Other
Rational Numbers

© 2019 Great Minds®. eureka-math.org

EUREKA MATH

b. Ezekiel drew two more cards, and his new score is the opposite of his original score. What two cards might he have drawn?

His cards must have a sum of 3.

The cards could be 1 and 5.

$$-7 + 4 + 1 + 5 = -7 + 10 = 3$$

I know that the new sum is the opposite of -3. That means that all four cards together must have a sum of 3. On a number line, I see that -3 and 3 are 6 units apart. So the two new cards must have a sum of 6.

4. $\frac{1}{5} + \left(-3\frac{7}{10}\right)$

To add fractions, I need common denominators. I will use the least common multiple of 5 and 10.

$$\frac{2}{10} + \left(-3\frac{7}{10}\right)$$

$$\frac{2}{10} + \left(-\frac{37}{10}\right)$$

I can write a mixed number as a fraction greater than 1 to help me add the numerators correctly.

$$-\frac{35}{10}$$

$$-3\frac{5}{10}$$

$$-3\frac{1}{2}$$

1. Find the sum. Show your work to justify your answer.

   a. $4 + 17$

   b. $-6 + (-12)$

   c. $2.2 + (-3.7)$

   d. $-3 + (-5) + 8$

   e. $\frac{1}{3} + \left(-2\frac{1}{4}\right)$

2. Which of these story problems describes the sum $19 + (-12)$? Check all that apply. Show your work to justify your answer.

   _____ Jared's dad paid him $19 for raking the leaves from the yard on Wednesday. Jared spent $12 at the movie theater on Friday. How much money does Jared have left?

   _____ Jared owed his brother $19 for raking the leaves while Jared was sick. Jared's dad gave him $12 for doing his chores for the week. How much money does Jared have now?

   _____ Jared's grandmother gave him $19 for his birthday. He bought $8 worth of candy and spent another $4 on a new comic book. How much money does Jared have left over?

3. Use the diagram below to complete each part.

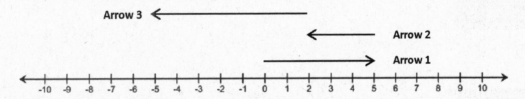

   a. Label each arrow with the number the arrow represents.

   b. How long is each arrow? What direction does each arrow point?

Arrow	Length	Direction
1		
2		
3		

   c. Write an equation that represents the sum of the numbers. Find the sum.

4. Jennifer and Katie were playing the Integer Game in class. Their hands are represented below.

Jennifer's Hand

| 5 | −8 |

Katie's Hand

| −9 | 7 |

a. What is the value of each of their hands? Show your work to support your answer.

b. If Jennifer drew two more cards, is it possible for the value of her hand not to change? Explain why or why not.

c. If Katie wanted to win the game by getting a score of 0, what card would she need? Explain.

d. If Jennifer drew −1 and −2, what would be her new score? Show your work to support your answer.

Lesson 4:   Efficiently Adding Integers and Other
Rational Numbers

EUREKA
MATH®

Example 1: Exploring Subtraction with the Integer Game

Play the Integer Game in your group. Start Round 1 by selecting four cards. Follow the steps for each round of play.

1. Write the value of your hand in the Total column.
2. Then, record what card values you select in the Action 1 column and discard from your hand in the Action 2 column.
3. After each action, calculate your new total, and record it under the appropriate Results column.
4. Based on the results, describe what happens to the value of your hand under the appropriate Descriptions column. For example, "Score increased by 3."

Round	Total	Action 1	Result 1	Description	Action 2	Result 2	Description
1							
2							
3							
4							
5							

Discussion: Making Connections to Integer Subtraction

1. How did selecting positive value cards change the value of your hand?

2. How did selecting negative value cards change the value of your hand?

3. How did discarding positive value cards change the value of your hand?

4. How did discarding negative value cards change the value of your hand?

5. What operation reflects selecting a card?

6. What operation reflects discarding or removing a card?

7. Based on the game, can you make a prediction about what happens to the result when

    a. Subtracting a positive integer?

    b. Subtracting a negative integer?

At the end of the lesson, the class reviews its predictions.

Lesson 5:   Understanding Subtraction of Integers and Other
                                Rational Numbers
                        © 2019 Great Minds®. eureka-math.org

EUREKA
MATH

Follow along with your teacher to complete the diagrams below.

$4 + 2 =$ ☐

Show that discarding (subtracting) a positive card, which is the same as subtracting a positive number, decreases the value of the hand.

$4 + 2 - 2 =$ ☐

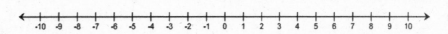

OR

$4 + 2 + (-2) =$ ☐

Removing (_____) a positive card changes the score in the same way as _____a card whose value

is the _____ _____ (or opposite). In this case, adding the corresponding

_____.

Example 3: Subtracting a Negative Number

Follow along with your teacher to complete the diagrams below.

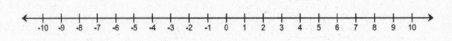

$4 + (-2) = \boxed{}$

How does removing a negative card change the score, or value, of the hand?

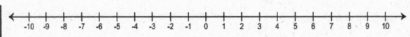

$4 + (-2) - (-2) = \boxed{}$

OR

$4 + (-2) + 2 = \boxed{}$

Removing (_____) a negative card changes the score in the same way as _____ a card whose value is the _____ _____ (or opposite). In this case, adding the corresponding _____.

EUREKA MATH

THE RULE OF SUBTRACTION: *Subtracting a number is the same as adding its additive inverse (or opposite).*

Exercises 1–3: Subtracting Positive and Negative Integers

1. Using the rule of subtraction, rewrite the following subtraction sentences as addition sentences and solve. Use the number line below if needed.

   a. $8 - 2$

   b. $4 - 9$

   c. $-3 - 7$

   d. $-9 - (-2)$

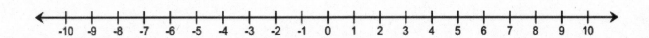

2. Find the differences.

   a. $-2 - (-5)$

   b. $11 - (-8)$

   c. $-10 - (-4)$

3. Write two equivalent expressions that represent the situation. What is the difference in their elevations?
   An airplane flies at an altitude of 25,000 feet. A submarine dives to a depth of 600 feet below sea level.

Lesson Summary

- **THE RULE OF SUBTRACTION:** Subtracting a number is the same as adding its opposite.

- Removing (subtracting) a positive card changes the score in the same way as adding a corresponding negative card.

- Removing (subtracting) a negative card makes the same change as adding the corresponding positive card.

- For all rational numbers, subtracting a number and adding it back gets you back to where you started: $(m - n) + n = m$.

EUREKA
MATH

Name _____   Date _____

1. If a player had the following cards, what is the value of his hand?

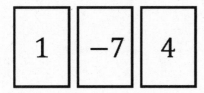

a. Identify two different ways the player could get to a score of 5 by adding or removing only one card.   Explain.

b. Write two equations for part (a), one for each of the methods you came up with for arriving at a score of 5.

2. Using the rule of subtraction, rewrite the following subtraction expressions as addition expressions, and find the sums.

a. $5 - 9$

b. $-14 - (-2)$

Adding the Opposite

1. Choose an integer between 3 and -3 on the number line, and label it point M. Locate and label the following points on the number line. Show your work.

I can choose any number I want between 3 and -3. For this example, I choose -2.

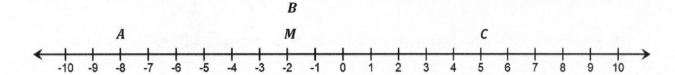

a. Point A: $M - 6$

   Point A : $-2 - 6$

   $\quad\quad\quad -2 + (-6)$

   $\quad\quad\quad -8$

   To subtract, I add the opposite. Therefore, I move to the left, just like I would if I subtract a positive.

b. Point B: $(M - 5) + 5$

   Point B : $(-2 - 5) + 5$

   $\quad\quad\quad (-2 + (-5)) + 5$

   $\quad\quad\quad -2$ *(same value as M)*

   If I subtract a number and then add the same number, the original value will remain the same. I can see this on the number line, moving left and then moving right the same number of units.

c. Point C: $-M - (-3)$

   Point C : $-(-2) - (-3)$

   $\quad\quad\quad 2 + 3$

   $\quad\quad\quad 5$

   When I see a negative in front of a number, I can read that as "the opposite of." So $-(-2)$ would mean the opposite of -2, which would be 2.

2. You and your partner were playing the Integer Game in class. Here are the cards in both hands.

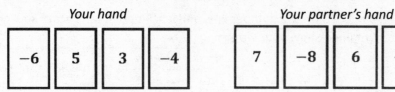

Your hand
| −6 | 5 | 3 | −4 |

Your partner's hand
| 7 | −8 | 6 | −3 |

a. Find the value of each hand. Who would win based on the current scores? (The score closest to 0 wins.)

> I find that both numbers are 2 units from 0, one is to the right, and the other is to the left. Neither player is closer to 0.

My hand: $-6 + 5 + 3 + (-4) = -2$

Partner's hand: $7 + (-8) + 6 + (-3) = 2$

My partner and I would tie because -2 and 2 are both 2 units from 0.

b. Find the value of each hand if you discarded the −4 and selected a 4 and your partner discarded the −3 and selected a 3. Show your work to support your answer. Then decide who would now win the game.

My hand: Discard the −4,

> Discarded means that it was thrown out or taken away. So I need to subtract those values from the scores I found in part (a).

$$-2 - (-4)$$
$$-2 + 4$$
$$2$$

Select a 4,

$$2 + 4$$
$$6$$

Partner's hand: Discard the − 3,

$$2 - (-3)$$
$$2 + 3$$
$$5$$

Select a 3,

$$5 + 3$$
$$8$$

My hand would win because 6 is the value that is closer to 0.

EUREKA MATH®

3. Explain what is meant by the following, and illustrate with an example:
   "For any real number, g, $4 - g = 4 + (-g)$."

> For this question, I need to explain why $4 - g$ is equivalent to $4 + (-g)$, and I have to give an example where I pick the value of g and show this is true.

Subtracting a number is the same as adding its additive inverse. Here is an example.

$g = 10$, $4 - (10)$ **is the same as** $4 + (-10)$ **because** -10 **is the opposite of** 10.

$4 - 10 = -6$

$4 + (-10) = -6$

So, $4 - 10 = 4 + (-10)$ **because they both equal** -6.

> I can substitute 10 for g in both expressions. Both expressions will give me the same answer (-6), showing the statement is true.

4. Write two equivalent expressions that represent the situation. What is the difference in their elevations?

   A mountain climber hikes to an altitude of 8,400 feet. A diver reaches a depth of 180 feet below sea level.

> A distance above sea level would be represented as a positive integer, and the distance below sea level would be a negative integer.

$$8,400 - (-180)$$
$$8,400 + 180$$
$$8,580$$

> I can also show this as addition. I need to add the distance above sea level with the distance below sea level to determine the total difference in their elevations.

The difference in their elevations is $8,580$ **ft.**

1. On a number line, find the difference of each number and 4. Complete the table to support your answers. The first example is provided.

Number	Subtraction Expression	Addition Expression	Answer
10	$10 - 4$	$10 + (-4)$	6
2			
-4			
-6			
1			

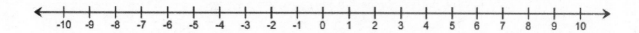

2. You and your partner were playing the Integer Game in class. Here are the cards in both hands.

Your hand

| −8 | 6 | 1 | −2 |

Your partner's hand

| 9 | −5 | 2 | −7 |

a. Find the value of each hand. Who would win based on the current scores? (The score closest to 0 wins.)

b. Find the value of each hand if you discarded the − 2 and selected a 5, and your partner discarded the − 5 and selected a 5. Show your work to support your answer.

c. Use your score values from part (b) to determine who would win the game now.

3. Write the following expressions as a single integer.
   a. $-2 + 16$
   b. $-2 - (-16)$
   c. $18 - 26$
   d. $-14 - 23$
   e. $30 - (-45)$

4. Explain what is meant by the following, and illustrate with an example:
   "For any real numbers, p and q, $p - q = p + (-q)$."

5. Choose an integer between -1 and -5 on the number line, and label it point P. Locate and label the following points on the number line. Show your work.

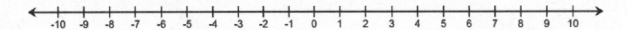

a. Point A: $P - 5$
b. Point B: $(P - 4) + 4$
c. Point C: $-P - (-7)$

Challenge Problem:

6. Write two equivalent expressions that represent the situation. What is the difference in their elevations?
   An airplane flies at an altitude of 26,000 feet. A submarine dives to a depth of 700 feet below sea level.

Lesson 5:   Understanding Subtraction of Integers and Other
             Rational Numbers

EUREKA
MATH®

Exercise 1

Use the number line to answer each of the following.

Person A	Person B
What is the distance between −4 and 5?	What is the distance between 5 and −4?
What is the distance between −5 and −3?	What is the distance between −3 and −5?
What is the distance between 7 and −1?	What is the distance between −1 and 7?

Exercise 2

Use the number line to answer each of the following questions.

a.   What is the distance between 0 and −8?

b.   What is the distance between −2 and $-1\frac{1}{2}$?

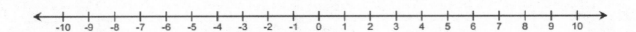

c.   What is the distance between −6 and −10?

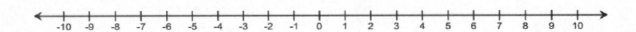

Example 1: Formula for the Distance Between Two Rational Numbers

Find the distance between −3 and 2.

Step 1: Start on an endpoint.

Step 2: Count the number of units from the endpoint you started on to the other endpoint.

Using a formula, _____

For two rational numbers p and q, the distance between p and q is $|p - q|$.

EUREKA
MATH®

Example 2: Change in Elevation vs. Distance

Distance is positive. Change in elevation or temperature may be positive or negative depending on whether it is increasing or decreasing (going up or down).

   a.   A hiker starts hiking at the beginning of a trail at a point which is 200 feet below sea level. He hikes to a location on the trail that is 580 feet above sea level and stops for lunch.

       i.   What is the vertical distance between 200 feet below sea level and 580 feet above sea level?

       ii.   How should we interpret 780 feet in the context of this problem?

   b.   After lunch, the hiker hiked back down the trail from the point of elevation, which is 580 feet above sea level, to the beginning of the trail, which is 200 feet below sea level.

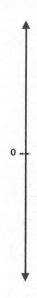

0

       i.   What is the vertical distance between 580 feet above sea level and 200 feet below sea level?

ii.   What is the change in elevation?

Exercise 3

The distance between a negative number and a positive number is $12\frac{1}{2}$. What are the numbers?

Exercise 4

Use the distance formula to find each answer. Support your answer using a number line diagram.

a.   Find the distance between -7 and -4.

b.   Find the change in temperature if the temperature rises from $-18°F$ to $15°F$ (use a vertical number line).

c. Would your answer for part (b) be different if the temperature dropped from 15°F to −18°F? Explain.

d. Beryl is the first person to finish a 5K race and is standing 15 feet beyond the finish line. Another runner, Jeremy, is currently trying to finish the race and has approximately 14 feet before he reaches the finish line. What is the minimum possible distance between Beryl and Jeremy?

e. What is the change in elevation from 140 feet above sea level to 40 feet below sea level? Explain.

> **Lesson Summary**
>
> - To find the distance between two rational numbers on a number line, you can count the number of units between the numbers.
> - Using a formula, the distance between rational numbers, p and q, is $|p - q|$.
> - Distance is always positive.
> - Change may be positive or negative. For instance, there is a $-4°$ change when the temperature goes from $7°$ to $3°$.

EUREKA MATH®

Name _____   Date _____

Two Grade 7 students, Monique and Matt, both solved the following math problem:

If the temperature drops from 7°F to −17°F, by how much did the temperature *decrease?*

The students came up with different answers. Monique said the answer is 24°F, and Matt said the answer is 10°F. Who is correct? Explain, and support your written response with the use of a formula and a vertical number line diagram.

1. Find the distance between the two rational numbers.

   a. $|-6-15|$

   $$|-6-15|$$
   $$|-6+(-15)|$$
   $$|-21|$$
   $$21$$

   > The bars on either side indicate absolute value. After I calculate the difference, I determine the absolute value. I can determine this by finding the distance the answer is from 0. Distances are always positive.

   b. $|6-(-15)|$

   $$|6-(-15)|$$
   $$|6+15|$$
   $$|21|$$
   $$21$$

   c. $|-7-5.4|$

   $$|-7-5.4|$$
   $$|-7+(-5.4)|$$
   $$|-12.4|$$
   $$12.4$$

   > I can work with signed decimals the same as I work with integers. I will still subtract by adding the opposite.

   d. $|7-(-5.4)|$

   $$|7-(-5.4)|$$
   $$|7+5.4|$$
   $$|12.4|$$
   $$12.4$$

2. Do you notice any special relationships between parts (a) and (b) or between parts (c) and (d)? Explain.

> The distance between two sets of opposites is the same.

The answers in parts (a) and (b) were the same because I was working with opposites. I found the distance between −6 to 15 and then found the distance between 6 and −15. The same relationship occurred in parts (c) and (d).

Lesson 6:   The Distance Between Two Rational Numbers

1. |−19 − 12|

2. |19 − (−12)|

3. |10 − (−43)|

4. |− 10 − 43|

5. |−1 − (−16)|

6. |1 − 16|

7. |0 − (−9)|

8. |0 − 9|

9. |− 14.5 − 13|

10. |14.5 − (−13)|

11. Describe any patterns you see in the answers to the problems in the left- and right-hand columns. Why do you think this pattern exists?

Exercise 1: Real-World Connection to Adding and Subtracting Rational Numbers

Suppose a seventh grader's birthday is today, and she is 12 years old. How old was she $3\frac{1}{2}$ years ago? Write an equation, and use a number line to model your answer.

Example 1: Representing Sums of Rational Numbers on a Number Line

a.   Place the tail of the arrow on 12.

b.   The length of the arrow is the absolute value of $-3\frac{1}{2}$, $\left|-3\frac{1}{2}\right| = 3\frac{1}{2}$.

c.   The direction of the arrow is to the *left* since you are adding a negative number to 12.

Draw the number line model in the space below.

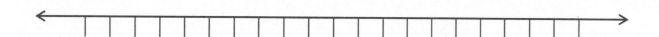

EUREKA
MATH®

© 2019 Great Minds®. eureka-math.org

Exercise 2

Find the following sum using a number line diagram: $-2\frac{1}{2} + 5$.

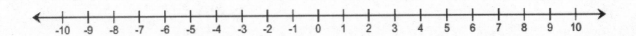

Example 2: Representing Differences of Rational Numbers on a Number Line

Find the following difference, and represent it on a number line: $1 - 2\frac{1}{4}$.

    a.

Now follow the steps to represent the sum:

    b.

    c.

    d.

Draw the number line model in the space below.

EUREKA
MATH®

Exercise 3

Find the following difference, and represent it on a number line: $-5\frac{1}{2} - (-8)$.

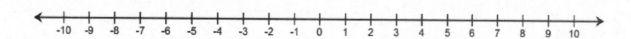

Exercise 4

Find the following sums and differences using a number line model.

a. $-6 + 5\frac{1}{4}$

b. $7 - (-0.9)$

c. $2.5 + \left(-\frac{1}{2}\right)$

d. $-\frac{1}{4} + 4$

e. $\frac{1}{2} - (-3)$

Exercise 5

Create an equation and number line diagram to model each answer.

a.   Samantha owes her father $7. She just got paid $5.50 for babysitting. If she gives that money to her dad, how much will she still owe him?

b.   At the start of a trip, a car's gas tank contains 12 gallons of gasoline. During the trip, the car consumes $10\frac{1}{8}$ gallons of gasoline. How much gasoline is left in the tank?

c.   A fish was swimming $3\frac{1}{2}$ feet below the water's surface at 7:00 a.m. Four hours later, the fish was at a depth that is $5\frac{1}{4}$ feet below where it was at 7:00 a.m. What rational number represents the position of the fish with respect to the water's surface at 11:00 a.m.?

Lesson 7:   Addition and Subtraction of Rational Numbers

EUREKA MATH®

Lesson Summary

The rules for adding and subtracting integers apply to all rational numbers.

The sum of two rational numbers (e.g., $-1 + 4.3$) can be found on the number line by placing the tail of an arrow at -1 and locating the head of the arrow 4.3 units to the right to arrive at the sum, which is 3.3.

To model the difference of two rational numbers on a number line (e.g., $-5.7 - 3$), first rewrite the difference as a sum, $-5.7 + (-3)$, and then follow the steps for locating a sum. Place a single arrow with its tail at -5.7 and the head of the arrow 3 units to the left to arrive at -8.7.

Name _____   Date _____

At the beginning of the summer, the water level of a pond is 2 feet below its normal level. After an unusually dry summer, the water level of the pond dropped another $1\frac{1}{3}$ feet.

1. Use a number line diagram to model the pond's current water level in relation to its normal water level.

2. Write an equation to show how far above or below the normal water level the pond is at the end of the summer.

Represent each of the following problems using both a number line diagram and an equation.

1. Mannah is going diving to check out different sea creatures. The total dive is 5.8 meters below sea level, and Mannah stops at 1.2 meters from the deepest part of the dive to look at a fish. How far from sea level will he be when he stops?

$$-5.8 - (-1.2)$$

$$-5.8 + 1.2$$

$$-4.6$$

> Because Mannah is below sea level, the depth of the dive will be negative. Then I will take away -1.2 meters because he already made this part of the journey back up to sea level (0 meters) before he stops to see the fish.

Mannah will be 4.6 meters below sea level when he stops to look at the fish.

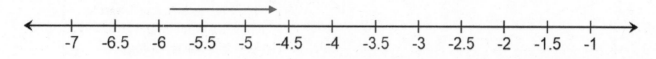

> I can check my work using the number line. I know that Mannah is heading towards sea level, or 0, so an arrow that is 1.2 units long facing right matches my equation and shows my work is correct.

2. A sturgeon was swimming $1\frac{1}{2}$ feet below sea level when it jumped up 4 feet before returning back to the water. How far above sea level was the fish at its highest point?

$$-1\frac{1}{2} + 4 = 2\frac{1}{2}$$

The sturgeon reached $2\frac{1}{2}$ feet above sea level.

> The initial location of the sturgeon is negative because it is below sea level. The sturgeon is jumping up, adding to its elevation.

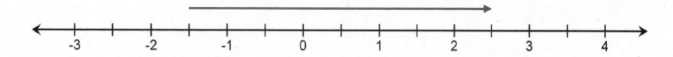

3. Marissa earned $16.75 babysitting and placed the money on a debit card. While shopping, she wanted to spend $22.40 on a new skirt. What would her new balance be on the debit card if she makes the purchase?

$$16.75 + (-22.40) = -5.65$$

The account balance would be $-$ \$5.65.

> The skirt Marissa wants to buy costs more than she earned. So she would have a negative balance on her debit card.

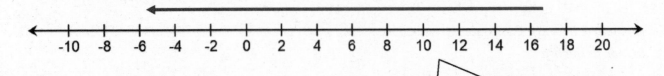

> This time my arrow will start at a positive number and go left to show that she is adding a negative when spending the money she earned.

Lesson 7:   Addition and Subtraction of Rational Numbers

EUREKA MATH®

Represent each of the following problems using both a number line diagram and an equation.

1. A bird that was perched atop a $15\frac{1}{2}$-foot tree dives down six feet to a branch below. How far above the ground is the bird's new location?

2. Mariah owed her grandfather $2.25 but was recently able to pay him back $1.50. How much does Mariah currently owe her grandfather?

3. Jake is hiking a trail that leads to the top of a canyon. The trail is 4.2 miles long, and Jake plans to stop for lunch after he completes 1.6 miles. How far from the top of the canyon will Jake be when he stops for lunch?

4. Sonji and her friend Rachel are competing in a running race. When Sonji is 0.4 miles from the finish line, she notices that her friend Rachel has fallen. If Sonji runs one-tenth of a mile back to help her friend, how far will she be from the finish line?

5. Mr. Henderson did not realize his checking account had a balance of $200 when he used his debit card for a $317.25 purchase. What is his checking account balance after the purchase?

6. If the temperature is −3°F at 10:00 p.m., and the temperature falls four degrees overnight, what is the resulting temperature?

Number Correct: _____

Integer Addition—Round 1

Directions: Determine the sum of the integers, and write it in the column to the right.

1.	$8 + (-5)$		18.	$-38 + 25$	
2.	$10 + (-3)$		19.	$-19 + (-11)$	
3.	$2 + (-7)$		20.	$2 + (-7)$	
4.	$4 + (-11)$		21.	$-23 + (-23)$	
5.	$-3 + (-9)$		22.	$45 + (-32)$	
6.	$-12 + (-7)$		23.	$16 + (-24)$	
7.	$-13 + 5$		24.	$-28 + 13$	
8.	$-4 + 9$		25.	$-15 + 15$	
9.	$7 + (-7)$		26.	$12 + (-19)$	
10.	$-13 + 13$		27.	$-24 + (-32)$	
11.	$14 + (-20)$		28.	$-18 + (-18)$	
12.	$6 + (-4)$		29.	$14 + (-26)$	
13.	$10 + (-7)$		30.	$-7 + 8 + (-3)$	
14.	$-16 + 9$		31.	$2 + (-15) + 4$	
15.	$-10 + 34$		32.	$-8 + (-19) + (-11)$	
16.	$-20 + (-5)$		33.	$15 + (-12) + 7$	
17.	$-18 + 15$		34.	$-28 + 7 + (-7)$	

Number Correct: _____
Improvement: _____

Integer Addition—Round 2

Directions: Determine the sum of the integers, and write it in the column to the right.

1.	$5 + (-12)$		18.	$23 + (-31)$	
2.	$10 + (-6)$		19.	$-26 + (-19)$	
3.	$-9 + (-13)$		20.	$16 + (-37)$	
4.	$-12 + 17$		21.	$-21 + 14$	
5.	$-15 + 15$		22.	$33 + (-8)$	
6.	$16 + (-25)$		23.	$-31 + (-13)$	
7.	$-12 + (-8)$		24.	$-16 + 16$	
8.	$-25 + (-29)$		25.	$30 + (-43)$	
9.	$28 + (-12)$		26.	$-22 + (-18)$	
10.	$-19 + (-19)$		27.	$-43 + 27$	
11.	$-17 + 20$		28.	$38 + (-19)$	
12.	$8 + (-18)$		29.	$-13 + (-13)$	
13.	$13 + (-15)$		30.	$5 + (-8) + (-3)$	
14.	$-10 + (-16)$		31.	$6 + (-11) + 14$	
15.	$35 + (-35)$		32.	$-17 + 5 + 19$	
16.	$9 + (-14)$		33.	$-16 + (-4) + (-7)$	
17.	$-34 + (-27)$		34.	$8 + (-24) + 12$	

Example 1: The Opposite of a Sum is the Sum of its Opposites

Explain the meaning of "The opposite of a sum is the sum of its opposites" Use a specific math example.

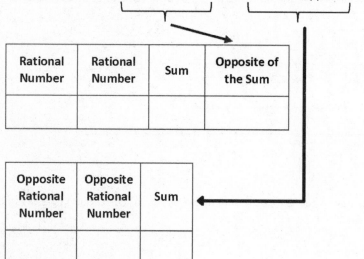

Rational Number	Rational Number	Sum	Opposite of the Sum

Opposite Rational Number	Opposite Rational Number	Sum

Exercise 1

Represent the following expression with a single rational number.

$$-2\frac{2}{5} + 3\frac{1}{4} - \frac{3}{5}$$

Example 2: A Mixed Number Is a Sum

Use the number line model shown below to explain and write the opposite of $2\frac{2}{5}$ as a sum of two rational numbers.

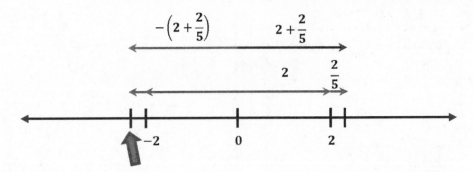

The opposite of a sum (top single arrow pointing left) and the sum of the opposites correspond to the same point on the number line.

Exercise 2

Rewrite each mixed number as the sum of two signed numbers.

a. $-9\frac{5}{8}$

b. $-2\frac{1}{2}$

c. $8\frac{11}{12}$

Exercise 3

Represent each sum as a mixed number.

a. $-1+\left(-\frac{5}{12}\right)$

b. $30+\frac{1}{8}$

c. $-17+\left(-\frac{1}{9}\right)$

EUREKA
MATH®

Exercise 4

Mr. Mitchell lost 10 pounds over the summer by jogging each week. By winter, he had gained $5\frac{1}{8}$ pounds. Represent this situation with an expression involving signed numbers. What is the overall change in Mr. Mitchell's weight?

Exercise 5

Jamal is completing a math problem and represents the expression $-5\frac{5}{7} + 8 - 3\frac{2}{7}$ with a single rational number as shown in the steps below. Justify each of Jamal's steps. Then, show another way to solve the problem.

$$= -5\frac{5}{7} + 8 + \left(-3\frac{2}{7}\right)$$

$$= -5\frac{5}{7} + \left(-3\frac{2}{7}\right) + 8$$

$$= -5 + \left(-\frac{5}{7}\right) + (-3) + \left(-\frac{2}{7}\right) + 8$$

$$= -5 + \left(-\frac{5}{7}\right) + \left(-\frac{2}{7}\right) + (-3) + 8$$

$$= -5 + (-1) + (-3) + 8$$

$$= -6 + (-3) + 8$$

$$= (-9) + 8$$

$$= -1$$

Lesson Summary

- Use the properties of operations to add and subtract rational numbers more efficiently. For instance,

$$-5\frac{2}{9}+3.7+5\frac{2}{9}=\left(-5\frac{2}{9}+5\frac{2}{9}\right)+3.7=0+3.7=3.7$$

- The opposite of a sum is the sum of its opposites as shown in the examples that follow:

$$-4\frac{4}{7}=-4+\left(-\frac{4}{7}\right)$$
$$-(5+3)=-5+(-3)$$

Lesson 8:   Applying the Properties of Operations to Add and
Subtract Rational Numbers

EUREKA
MATH

Name _____   Date _____

Mariah and Shane both started to work on a math problem and were comparing their work in math class. Are both of their representations correct? Explain, and finish the math problem correctly to arrive at the correct answer.

Math Problem

Jessica's friend lent her $5. Later that day Jessica gave her friend back $1\frac{3}{4}$ dollars.

Which rational number represents the overall change to the amount of money Jessica's friend has?

Mariah started the problem as follows:

$$-5 - \left(-1\frac{3}{4}\right)$$

$$= -5 + 1 - \frac{3}{4}$$

Shane started the problem as follows:

$$-5 - \left(-1\frac{3}{4}\right)$$

$$= -5 + \left(1\frac{3}{4}\right)$$

$$= -5 + \left(1 + \frac{3}{4}\right)$$

_____

_____

_____

_____

_____

1. Jerod dropped his wallet at the grocery store. The wallet contained $40. When he got home, his grandfather felt sorry for him and gave him $28.35. Represent this situation with an expression involving rational numbers. What is the overall change in the amount of money Jerod has?

$$-40 + 28.35$$

$$-11.65$$

> Losing the wallet would be a negative because now he doesn't have the money anymore. But then we need to add a positive on to the total when the grandfather gives him money.

The overall change in the amount of money Jerod has is −11.65 **dollars.**

> The sum of losing the money and then gaining money should be closer to zero but still negative because Jerod was given less money than he lost.

2. Zoe is completing some math problems. What are the answers? Show your work.

a. $-9 + -\dfrac{2}{3}$

> In the lesson, I practiced writing a mixed number as the sum of two signed numbers. I can do the reverse of that here and write the sum of two signed numbers as a mixed number.

$$-9\dfrac{2}{3}$$

b. $16 - 19\dfrac{4}{5}$

$$16 + \left(-19\dfrac{4}{5}\right)$$

> I write the mixed number as the sum of two signed numbers, and then I can combine the integers together. Finally, the two signed numbers will combine to form a mixed number.

$$16 + \left(-19 + \left(-\dfrac{4}{5}\right)\right)$$

$$16 + (-19) + \left(-\dfrac{4}{5}\right)$$

$$-3 + \left(-\dfrac{4}{5}\right)$$

$$-3\dfrac{4}{5}$$

c. $\left(\dfrac{1}{8}+\dfrac{3}{4}\right)+\left(\left(-\dfrac{1}{8}\right)+\left(-\dfrac{3}{4}\right)\right)$

$\left(\dfrac{1}{8}+\dfrac{6}{8}\right)+\left(\left(-\dfrac{1}{8}\right)+\left(-\dfrac{6}{8}\right)\right)$

> I rewrite the fractions with common denominators before adding.

$\left(\dfrac{7}{8}\right)+\left(-\dfrac{7}{8}\right)$

0

> I am adding a sum and its opposite. I know this because the second sum in parentheses has the opposite sign but the same absolute value.

Lesson 8:   Applying the Properties of Operations to Add and
               Subtract Rational Numbers

EUREKA
MATH®

1. Represent each sum as a single rational number.

   a. $-14 + \left(-\frac{8}{9}\right)$

   b. $7 + \frac{1}{9}$

   c. $-3 + \left(-\frac{1}{6}\right)$

Rewrite each of the following to show that *the opposite of a sum is the sum of the opposites*. Problem 2 has been completed as an example.

2. $-(9+8) = -9 + (-8)$
   $-17 = -17$

3. $-\left(\frac{1}{4} + 6\right)$

4. $-(10 + (-6))$

5. $-\left((-55) + \frac{1}{2}\right)$

Use your knowledge of rational numbers to answer the following questions.

6. Meghan said the opposite of the sum of -12 and 4 is 8. Do you agree? Why or why not?

7. Jolene lost her wallet at the mall. It had $10 in it. When she got home, her brother felt sorry for her and gave her $5.75. Represent this situation with an expression involving rational numbers. What is the overall change in the amount of money Jolene has?

8. Isaiah is completing a math problem and is at the last step: $25 - 28\frac{1}{5}$. What is the answer? Show your work.

9. A number added to its opposite equals zero. What do you suppose is true about *a sum added to its opposite*? Use the following examples to reach a conclusion. Express the answer to each example as a single rational number.

   a. $(3+4) + (-3 + -4)$
   b. $(-8+1) + (8 + (-1))$

   c. $\left(-\frac{1}{2} + \left(-\frac{1}{4}\right)\right) + \left(\frac{1}{2} + \frac{1}{4}\right)$

Number Correct: _____

Directions: Determine the difference of the integers, and write it in the column to the right.

1.	4 − 2		23.	(−6) − 5	
2.	4 − 3		24.	(−6) − 7	
3.	4 − 4		25.	(−6) − 9	
4.	4 − 5		26.	(−14) − 9	
5.	4 − 6		27.	(−25) − 9	
6.	4 − 9		28.	(−12) − 12	
7.	4 − 10		29.	(−26) − 26	
8.	4 − 20		30.	(−13) − 21	
9.	4 − 80		31.	(−25) − 75	
10.	4 − 100		32.	(−411) −811	
11.	4 − (−1)		33.	(−234) − 543	
12.	4 − (−2)		34.	(−3) − (−1)	
13.	4 − (−3)		35.	(−3) − (−2)	
14.	4 − (−7)		36.	(−3) − (−3)	
15.	4 − (−17)		37.	(−3) − (−4)	
16.	4 − (−27)		38.	(−3) − (−8)	
17.	4 − (−127)		39.	(−30) − (−45)	
18.	14 − (−6)		40.	(−27) − (−13)	
19.	23 − (−8)		41.	(−13) − (−27)	
20.	8 − (−23)		42.	(−4) − (−3)	
21.	51 − (−3)		43.	(−3) − (−4)	
22.	48 − (−5)		44.	(−1,066) − (−34)	

Number Correct: _____

Improvement: _____

Integer Subtraction—Round 2

Directions: Determine the difference of the integers, and write it in the column to the right.

1.	$3 - 2$		23.	$(-8) - 5$	
2.	$3 - 3$		24.	$(-8) - 7$	
3.	$3 - 4$		25.	$(-8) - 9$	
4.	$3 - 5$		26.	$(-15) - 9$	
5.	$3 - 6$		27.	$(-35) - 9$	
6.	$3 - 9$		28.	$(-22) - 22$	
7.	$3 - 10$		29.	$(-27) - 27$	
8.	$3 - 20$		30.	$(-14) - 21$	
9.	$3 - 80$		31.	$(-22) - 72$	
10.	$3 - 100$		32.	$(-311) - 611$	
11.	$3 - (-1)$		33.	$(-345) - 654$	
12.	$3 - (-2)$		34.	$(-2) - (-1)$	
13.	$3 - (-3)$		35.	$(-2) - (-2)$	
14.	$3 - (-7)$		36.	$(-2) - (-3)$	
15.	$3 - (-17)$		37.	$(-2) - (-4)$	
16.	$3 - (-27)$		38.	$(-2) - (-8)$	
17.	$3 - (-127)$		39.	$(-20) - (-45)$	
18.	$13 - (-6)$		40.	$(-24) - (-13)$	
19.	$24 - (-8)$		41.	$(-13) - (-24)$	
20.	$5 - (-23)$		42.	$(-5) - (-3)$	
21.	$61 - (-3)$		43.	$(-3) - (-5)$	
22.	$58 - (-5)$		44.	$(-1,034) - (-31)$	

EUREKA
MATH®

Lesson 9:   Applying the Properties of Operations to Add and
            Subtract Rational Numbers

105

© 2019 Great Minds®. eureka-math.org

Exercise 1

Unscramble the cards, and show the steps in the correct order to arrive at the solution to $5\frac{2}{9} - \left(8.1 + 5\frac{2}{9}\right)$.

$$0 + (-8.1)$$

$$\left(5\frac{2}{9} + \left(-5\frac{2}{9}\right)\right) + (-8.1)$$

$$-8.1$$

$$5\frac{2}{9} + \left(-8.1 + \left(-5\frac{2}{9}\right)\right)$$

$$5\frac{2}{9} + \left(-5\frac{2}{9} + (-8.1)\right)$$

Examples 1–2

Represent each of the following expressions as one rational number. Show and explain your steps.

1. $4\frac{4}{7} - \left(4\frac{4}{7} - 10\right)$

2. $5 + \left(-4\frac{4}{7}\right)$

EUREKA MATH

Exercise 2: Team Work!

a. $-5.2 - (-3.1) + 5.2$

b. $32 + \left(-12\frac{7}{8}\right)$

c. $3\frac{1}{6} + 20.3 - \left(-5\frac{5}{6}\right)$

d. $\frac{16}{20} - (-1.8) - \frac{4}{5}$

Exercise 3

Explain, step by step, how to arrive at a single rational number to represent the following expression. Show both a written explanation and the related math work for each step.

$$-24 - \left(-\frac{1}{2}\right) - 12.5$$

Lesson Summary

- Use the properties of operations to add and subtract rational numbers more efficiently. For instance,

$$-5\frac{2}{9} + 3.7 + 5\frac{2}{9} = \left(-5\frac{2}{9} + 5\frac{2}{9}\right) + 3.7 = 0 + 3.7 = 3.7$$

- The opposite of a sum is the sum of its opposites as shown in the examples that follow:

$$-4\frac{4}{7} = -4 + \left(-\frac{4}{7}\right)$$

$$-(5+3) = -5 + (-3).$$

Lesson 9: Applying the Properties of Operations to Add and Subtract Rational Numbers

EUREKA MATH

Name _____ Date _____

1. Jamie was working on his math homework with his friend, Kent. Jamie looked at the following problem.

$$-9.5 - (-8) - 6.5$$

He told Kent that he did not know how to subtract negative numbers. Kent said that he knew how to solve the problem using only addition. What did Kent mean by that? Explain. Then, show your work, and represent the answer as a single rational number.

_____

_____

_____

Work Space:

Answer: _____

2. Use one rational number to represent the following expression. Show your work.

$$3 + (-0.2) - 15\frac{1}{4}$$

1. Show all steps needed to rewrite each of the following expressions as a single rational number.

   a. $14 - \left(-8\frac{4}{9}\right)$

   $$14 + 8\frac{4}{9}$$

   $$14 + 8 + \frac{4}{9}$$

   $$22 + \frac{4}{9}$$

   $$22\frac{4}{9}$$

   > I can separate the mixed number so that I can work with the whole number and the fraction separately.

   b. $-2\frac{2}{5} + 4.1 - 8\frac{1}{5}$

   $$-2\frac{2}{5} - 8\frac{1}{5} + 4.1$$

   $$-2\frac{2}{5} + \left(-8\frac{1}{5}\right) + 4.1$$

   > I apply the commutative property because the two mixed numbers have common denominators already.

   $$-2 + \left(-\frac{2}{5}\right) + (-8) + \left(-\frac{1}{5}\right) + 4.1$$

   $$-2 + (-8) + \left(-\frac{2}{5}\right) + \left(-\frac{1}{5}\right) + 4.1$$

   $$-10 + -\frac{3}{5} + 4.1$$

   $$-10\frac{3}{5} + 4.1$$

   > I can rewrite the decimal as a fraction, $4\frac{1}{10}$. And I can write $\frac{3}{5}$ as $\frac{6}{10}$ so that I have common denominators to add.

   $$-10\frac{3}{5} + 4\frac{1}{10}$$

   $$-10\frac{6}{10} + 4\frac{1}{10}$$

   $$-6\frac{5}{10}$$

EUREKA MATH

2. Explain, step by step, how to arrive at a single rational number to represent the following expression. Show both a written explanation and the related math work for each step.

$$4 - \left(-3\frac{2}{9}\right) + 2\frac{1}{3}$$

Rewrite the subtraction as adding the inverse.	$4 + 3\frac{2}{9} + 2\frac{1}{3}$
Get common denominators.	$4 + 3\frac{2}{9} + 2\frac{3}{9}$
Separate each mixed number into the sum of its parts.	$4 + 3 + \frac{2}{9} + 2 + \frac{3}{9}$
Reorder the addends so that I can add the whole number addends and add the fractional addends.	$4 + 3 + 2 + \frac{2}{9} + \frac{3}{9}$
Add the whole number addends and then the fractional addends.	$9 + \frac{5}{9}$
Add the whole number addend and fractional addend together.	$9\frac{5}{9}$

Lesson 9: Applying the Properties of Operations to Add and
Subtract Rational Numbers

EUREKA
MATH®

Show all steps taken to rewrite each of the following as a single rational number.

1. $80 + \left(-22\frac{4}{15}\right)$

2. $10 + \left(-3\frac{3}{8}\right)$

3. $\frac{1}{5} + 20.3 - \left(-5\frac{3}{5}\right)$

4. $\frac{11}{12} - \left(-10\right) - \frac{5}{6}$

5. Explain, step by step, how to arrive at a single rational number to represent the following expression. Show both a written explanation and the related math work for each step.

$$1 - \frac{3}{4} + \left(-12\frac{1}{4}\right)$$

Exercise 1: Integer Game Revisited

In groups of four, play one round of the Integer Game (see Integer Game outline for directions).

Example 1: Product of a Positive Integer and a Negative Integer

Part A:

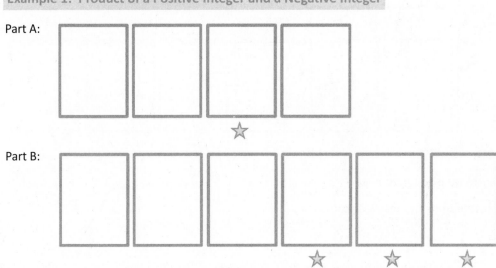

Part B:

Use your cards from Part B to answer the questions below.

a. Write a product that describes the three matching cards.

b. Write an expression that represents how each of the ✩ cards changes your score.

c. Write an equation that relates these two expressions.

d. Write an integer that represents the total change to your score by the three ✩ cards.

e. Write an equation that relates the product and how it affects your score.

Part C:

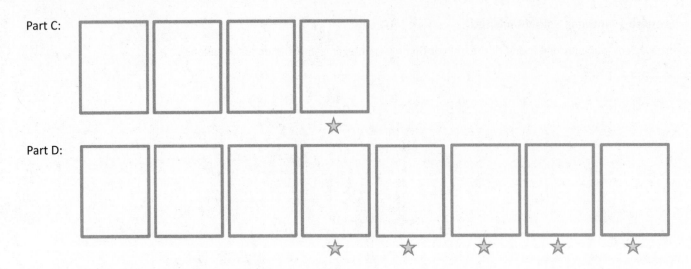

Part D:

Use your cards from Part D to answer the questions below.

f.   Write a product that describes the five matching cards.

g.   Write an expression that represents how each of the ☆ cards changes your score.

h.   Write an equation that relates these two expressions.

i.   Write an integer that represents the total change to your score by the five ☆ cards.

j.   Write an equation that relates the product and how it affects your score.

k.   Use the expression 5 × 4 to relate the multiplication of a positive valued card to addition.

l.   Use the expression 3 × (−5) to relate the multiplication of a negative valued card to addition.

EUREKA
MATH

Example 2: Product of a Negative Integer and a Positive Integer

a. If all of the 4's from the playing hand on the right are discarded, how will the score be affected? Model this using a product in an equation.

b. What three matching cards could be added to those pictured to get the same change in score? Model this using a product in an equation.

c. Seeing how each play affects the score, relate the products that you used to model them. What do you conclude about multiplying integers with opposite signs?

Example 3: Product of Two Negative Integers

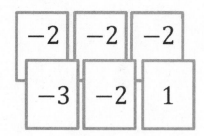

a. If the matching cards from the playing hand on the right are discarded, how will this hand's score be affected? Model this using a product in an equation.

b. What four matching cards could be added to those pictured to get the same change in score? Model this using a product in an equation.

EUREKA
MATH®

c. Seeing how each play affects the score, relate the products that you used to model them. What do you conclude about multiplying integers with the same sign?

d. Using the conclusions from Examples 2 and 3, what can we conclude about multiplying integers? Write a few examples.

EUREKA MATH®

Lesson Summary

Multiplying integers is repeated addition and can be modeled with the Integer Game. If $3 \times a$ corresponds to what happens to your score if you get three cards of value a, then $(-3) \times a$ corresponds to what happens to your score if you lose three cards of value a. Adding a number multiple times has the same effect as removing the opposite value the same number of times (e.g., $a \times b = (-a) \times (-b)$ and $a \times (-b) = (-a) \times b$).

Name _____   Date _____

1. Natalie is playing the Integer Game and only shows you the four cards shown below. She tells you that the rest of her cards have the same values on them and match one of these four cards.

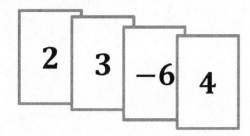

   a. If all of the matching cards will increase her score by 18, what are the matching cards?

   b. If all of the matching cards will decrease her score by 12, what are the matching cards?

2. A hand of six integer cards has one matching set of two or more cards. If the matching set of cards is removed from the hand, the score of the hand will increase by six. What are the possible values of these matching cards? Explain. Write an equation using multiplication showing how the matching cards yield an increase in score of six.

Integer Game

Describe sets of two or more matching integer cards that satisfy the criteria in each problem below.

1. Cards increase the score by six points.

    Picking up: six 1's, three 2's, or two 3's

    OR

    Removing: six (-1)'s, three (-2)'s, or two (-3)'s

    > I can increase my score two ways: picking up positive value cards (pos. × pos.) or removing negative value cards from my hand (neg. × neg.).

2. Cards decrease the score by 4 points.

    Picking up: four (-1)'s or two (-2)'s

    OR

    Removing: four 1's or two 2's

    > I can decrease my score two ways: picking up negative value cards (pos. × neg.) or removing positive value cards from my hand (neg. × pos.).

3. Removing cards that increase the score by 8 points.

    Eight (-1)'s, four (-2)'s, or two (-4)'s

    > If I want to remove cards to increase my score, I must remove negative value cards.

4. Bruce is playing the Integer Game and is given the opportunity to discard a set of matching cards. Bruce determines that if he discards one set of cards, his score will increase by 8. If he discards another set, then his score will decrease by 15. If his matching cards make up all seven cards in his hand, what cards are in Bruce's hand?

    There are two possibilities:

    −4, −4, 3, 3, 3, 3, 3 or −2, −2, −2, −2, 5, 5, 5

    > Removing negative cards increases Bruce's score and removing positive cards decreases Bruce's score.

1. Describe sets of two or more matching integer cards that satisfy the criteria in each part below:

   a. Cards increase the score by eight points.

   b. Cards decrease the score by 9 points.

   c. Removing cards that increase the score by 10 points.

   d. Positive cards that decrease the score by 18 points.

2. You have the integer cards shown at the right when your teacher tells you to choose a card to multiply four times. If your goal is to get your score as close to zero as possible, which card would you choose? Explain how your choice changes your score.

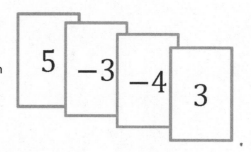

3. Sherry is playing the Integer Game and is given a chance to discard a set of matching cards. Sherry determines that if she discards one set of cards, her score will increase by 12. If she discards another set, then her score will decrease by eight. If her matching cards make up all six cards in her hand, what cards are in Sherry's hand? Are there any other possibilities?

Example 1: Extending Whole Number Multiplication to the Integers

Part A: Complete quadrants *I* and *IV* of the table below to show how sets of matching integer cards will affect a player's score in the Integer Game. For example, three 2s would increase a player's score by $0 + 2 + 2 + 2 = 6$ points.

Quadrant *II*   Quadrant *I*

What does this quadrant represent?

_____

_____

_____

_____

What does this quadrant represent?

_____

_____

					5					
					4					
					3					
					2		6			
					1					
						1	2	3	4	5
					−1					
					−2					
					−3					
					−4					
					−5					

← Number of matching cards

What does this quadrant represent?

_____

_____

_____

_____

What does this quadrant represent?

_____

_____

_____

_____

Quadrant *III*   ↑   Quadrant *IV*

Integer card values

a.   What patterns do you see in the right half of the table?

b.   Enter the missing integers in the left side of the middle row, and describe what they represent.

Part B: Complete quadrant *II* of the table.

    c.   What relationships or patterns do you notice between the products (values) in quadrant *II* and the products (values) in quadrant *I*?

    d.   What relationships or patterns do you notice between the products (values) in quadrant *II* and the products (values) in quadrant *IV*?

    e.   Use what you know about the products (values) in quadrants *I*, *II*, and *IV* to describe what quadrant *III* will look like when its products (values) are entered.

Part C: Complete quadrant *III* of the table.

Refer to the completed table to help you answer the following questions:

    f.   Is it possible to know the sign of a product of two integers just by knowing in which quadrant each integer is located? Explain.

    g.   Which quadrants contain which values? Describe an Integer Game scenario represented in each quadrant.

Lesson 11:   Develop Rules for Multiplying Signed Numbers

EUREKA MATH

Example 2: Using Properties of Arithmetic to Explain Multiplication of Negative Numbers

Exercise 1: Multiplication of Integers in the Real World

Generate real-world situations that can be modeled by each of the following multiplication problems. Use the Integer Game as a resource.

     a. -3×5

     b. $-6 \times (-3)$

     c. $4 \times (-7)$

Lesson Summary

To multiply signed numbers, multiply the absolute values to get the absolute value of the product. The sign of the product is positive if the factors have the same sign and negative if they have opposite signs.

Lesson 11: Develop Rules for Multiplying Signed Numbers

Name _____   Date _____

1. Create a real-life example that can be modeled by the expression -2×4, and then state the product.

2. Two integers are multiplied, and their product is a positive number. What must be true about the two integers?

Multiplying Integers

1. Explain why $(-3) \times (-2) = 6$. Use patterns, an example from the Integer Game, or the properties of operations to support your reasoning.

> Instead of using the Integer Game to explain why the product is positive, I could have also used the patterns I saw in the four quadrants during class today.

> *If I think about the Integer Game, removing negative cards from my hand increases my score. Therefore, the problem presented indicates that I removed three cards (-3) each with a value of -2. This change would increase my score by 6 points.*

2. Emilia receives allergy shots in order to decrease her allergy symptoms. Emilia must pay $20 each time she receives a shot, which is twice a week. Write an integer that represents the change in Emilia's money from receiving shots for 8 weeks. Explain your reasoning.

$$2(8) = 16$$

Emilia receives 16 shots in 8 weeks.

> Each time Emilia pays for a shot, her money decreases by $20, which is indicated with -20 in my equation.

$$-20(16) = -320$$

The change in Emilia's money after 8 weeks of receiving shots twice a week is $-\$320$.

> I could have picked a different real-world example, but it would still require that -2 is repeated three times.

3. Write a real-world problem that can be modeled by $3 \times (-2)$.

The temperature has decreased two degrees each hour for the last three hours. What is the change in temperature after the three hours?

> A decrease of 2 degrees describes the -2 in the original expression.

1. Complete the problems below. Then, answer the question that follows.

$3 \times 3 =$ $3 \times 2 =$ $3 \times 1 =$ $3 \times 0 =$ $3 \times (-1) =$ $3 \times (-2) =$

$2 \times 3 =$ $2 \times 2 =$ $2 \times 1 =$ $2 \times 0 =$ $2 \times (-1) =$ $2 \times (-2) =$

$1 \times 3 =$ $1 \times 2 =$ $1 \times 1 =$ $1 \times 0 =$ $1 \times (-1) =$ $1 \times (-2) =$

$0 \times 3 =$ $0 \times 2 =$ $0 \times 1 =$ $0 \times 0 =$ $0 \times (-1) =$ $0 \times (-2) =$

$-1 \times 3 =$ $-1 \times 2 =$ $-1 \times 1 =$ $-1 \times 0 =$ $-1 \times (-1) =$ $-1 \times (-2) =$

$-2 \times 3 =$ $-2 \times 2 =$ $-2 \times 1 =$ $-2 \times 0 =$ $-2 \times (-1) =$ $-2 \times (-2) =$

$-3 \times 3 =$ $-3 \times 2 =$ $-3 \times 1 =$ $-3 \times 0 =$ $-3 \times (-1) =$ $-3 \times (-2) =$

Which row shows the same pattern as the outlined column? Are the problems similar or different? Explain.

2. Explain why $(-4) \times (-5) = 20$. Use patterns, an example from the Integer Game, or the properties of operations to support your reasoning.

3. Each time that Samantha rides the commuter train, she spends $4 for her fare. Write an integer that represents the change in Samantha's money from riding the commuter train to and from work for 13 days. Explain your reasoning.

4. Write a real-world problem that can be modeled by $4 \times (-7)$.

Challenge:

5. Use properties to explain why for each integer a, $-a = -1 \times a$. (Hint: What does $(1 + (-1)) \times a$ equal? What is the additive inverse of a?)

Exercise 1: Recalling the Relationship Between Multiplication and Division

Record equations from Exercise 1 on the left.

Equations   **Integers**

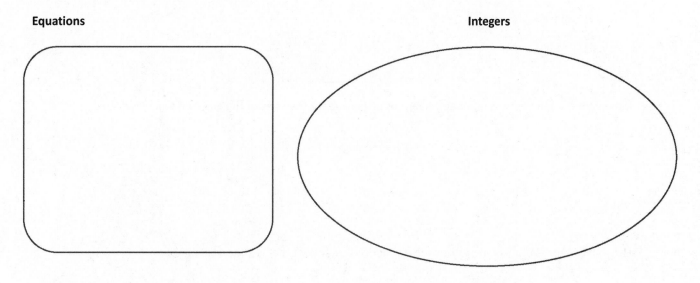

Record your group's number sentences in the space on the left below.

Equations   **Integers**

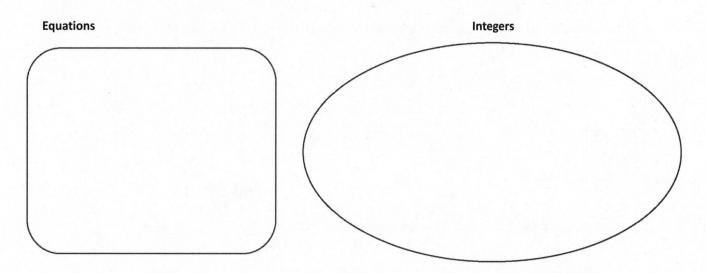

a. List examples of division problems that produced a quotient that is a negative number.

b. If the quotient is a negative number, what must be true about the signs of the dividend and divisor?

c. List your examples of division problems that produced a quotient that is a positive number.

d. If the quotient is a positive number, what must be true about the signs of the dividend and divisor?

Rules for Dividing Two Integers:

- A quotient is negative if the divisor and the dividend have _____ signs.

- A quotient is positive if the divisor and the dividend have _____ signs.

EUREKA MATH

Exercise 2: Is the Quotient of Two Integers Always an Integer?

Is the quotient of two integers always an integer? Use the work space below to create quotients of integers. Answer the question, and use examples or a counterexample to support your claim.

Work Space:

Answer:

Exercise 3: Different Representation of the Same Quotient

Are the answers to the three quotients below the same or different? Why or why not?

    a. $-14 \div 7$

    b. $14 \div (-7)$

    c. $-(14 \div 7)$

Lesson Summary

The rules for dividing integers are similar to the rules for multiplying integers (when the divisor is not zero). The quotient is positive if the divisor and dividend have the same signs and negative if they have opposite signs.

The quotient of any two integers (with a nonzero divisor) will be a rational number. If p and q are integers, then

$$-\left(\frac{p}{q}\right) = \frac{-p}{q} = \frac{p}{-q}.$$

EUREKA MATH

Fluency Exercise: Integer Division

1.	$-56 \div (-7) =$	15.	$-28 \div (-7) =$	29.	$-14 \div (-7) =$
2.	$-56 \div (-8) =$	16.	$-28 \div (-4) =$	30.	$-14 \div (-2) =$
3.	$56 \div (-8) =$	17.	$28 \div 4 =$	31.	$14 \div (-2) =$
4.	$-56 \div 7 =$	18.	$-28 \div 7 =$	32.	$-14 \div 7 =$
5.	$-40 \div (-5) =$	19.	$-20 \div (-5) =$	33.	$-10 \div (-5) =$
6.	$-40 \div (-4) =$	20.	$-20 \div (-4) =$	34.	$-10 \div (-2) =$
7.	$40 \div (-4) =$	21.	$20 \div (-4) =$	35.	$10 \div (-2) =$
8.	$-40 \div 5 =$	22.	$-20 \div 5 =$	36.	$-10 \div 5 =$
9.	$-16 \div (-4) =$	23.	$-8 \div (-4) =$	37.	$-4 \div (-4) =$
10.	$-16 \div (-2) =$	24.	$-8 \div (-2) =$	38.	$-4 \div (-1) =$
11.	$16 \div (-2) =$	25.	$8 \div (-2) =$	39.	$4 \div (-1) =$
12.	$-16 \div 4 =$	26.	$-8 \div 4 =$	40.	$-4 \div 1 =$
13.	$-3 \div (-4) =$	27.	$4 \div (-8) =$	41.	$1 \div (-4) =$
14.	$-3 \div 4 =$	28.	$-4 \div 8 =$	42.	$-1 \div 4 =$

Name _____ Date _____

1. Mrs. McIntire, a seventh-grade math teacher, is grading papers. Three students gave the following responses to the same math problem:

Student one: $\dfrac{1}{-2}$

Student two: $-\left(\dfrac{1}{2}\right)$

Student three: $-\dfrac{1}{2}$

On Mrs. McIntire's answer key for the assignment, the correct answer is −0.5. Which student answer(s) is (are) correct? Explain.

2. Complete the table below. Provide an answer for each integer division problem, and write a related equation using integer multiplication.

Integer Division Problem	Related Equation Using Integer Multiplication
−36 ÷ (−9) = _____	
24 ÷ (−8) = _____	
50 ÷ 10 = _____	
42 ÷ 6 = _____	

1. Find the missing value in each column.

Column A

$-16 \div 8 = \textbf{-2}$

$16 \div -8 = \textbf{-2}$

$-16 \div -8 = \textbf{2}$

$16 \div 8 = \textbf{2}$

Column B

$-32 \div 8 = \textbf{-4}$

$32 \div -8 = \textbf{-4}$

$-32 \div -8 = \textbf{4}$

$32 \div 8 = \textbf{4}$

> I know that when the dividend and divisor have the same sign, the quotient will be positive. I also know that when the dividend and divisor have opposite signs, the quotient will be negative.

2. Describe the pattern you see in each column's answers in Problem 1, relating it to the divisors and dividends. Why is this so?

The first two quotients in each column are negative because the dividend and divisor have opposite signs but the same absolute values, which means the answers will have a negative value. The last two quotients in each column are positive because the dividend and divisor have the same signs and absolute values, which means the answer will be a positive value.

3. Describe the pattern you see between the answers in Column A and Column B in Problem 1. Why is this so?

The quotients in Column B are each double the corresponding quotients in Column A. This is true because the divisors in each column are the same, but the dividends in Column B are double the corresponding dividends in Column A. Since 32 is double 16 and the divisors remain the same, the quotients will also be double.

> If the divisor's value doubled and the dividend remained constant, the value of the new quotient would be half of the value of the original quotient.

1. Find the missing values in each column.

Column A	Column B	Column C	Column D
$48 \div 4 =$	$24 \div 4 =$	$63 \div 7 =$	$21 \div 7 =$
$-48 \div (-4) =$	$-24 \div (-4) =$	$-63 \div (-7) =$	$-21 \div (-7) =$
$-48 \div 4 =$	$-24 \div 4 =$	$-63 \div 7 =$	$-21 \div 7 =$
$48 \div (-4) =$	$24 \div (-4) =$	$63 \div (-7) =$	$21 \div (-7) =$

2. Describe the pattern you see in each column's answers in Problem 1, relating it to the problems' divisors and dividends. Why is this so?

3. Describe the pattern you see between the answers for Columns A and B in Problem 1 (e.g., compare the first answer in Column A to the first answer in Column B; compare the second answer in Column A to the second answer in Column B). Why is this so?

4. Describe the pattern you see between the answers for Columns C and D in Problem 1. Why is this so?

Example 1: Representations of Rational Numbers in the Real World

Following the Opening Exercise and class discussion, describe why we need to know how to represent rational numbers in different ways.

Example 2: Using Place Values to Write (Terminating) Decimals as Equivalent Fractions

a. What is the value of the number 2.25? How can this number be written as a fraction or mixed number?

b. Rewrite the fraction in its simplest form showing all steps that you use.

c. What is the value of the number 2.025? How can this number be written as a mixed number?

d. Rewrite the fraction in its simplest form showing all steps that you use.

Lesson 13:   Converting Between Fractions and Decimals Using
               Equivalent Fractions

151

© 2019 Great Minds®. eureka-math.org

Exercise 1

Use place value to convert each terminating decimal to a fraction. Then rewrite each fraction in its simplest form.

  a. 0.218

  b. 0.16

  c. 2.72

  d. 0.0005

  a. What are *decimals*?

EUREKA
MATH

b. Use the meaning of *decimal* to relate decimal place values.

c. Write the number $\dfrac{3}{100}$ as a decimal. Describe your process.

d. Write the number $\dfrac{3}{20}$ as a decimal. Describe your process.

e. Write the number $\dfrac{10}{25}$ as a decimal. Describe your process.

f. Write the number $\dfrac{8}{40}$ as a decimal. Describe your process.

Exercise 2

Convert each fraction to a decimal using an equivalent fraction.

a. $\dfrac{3}{16} =$

b. $\dfrac{7}{5} =$

c. $\dfrac{11}{32} =$

d. $\dfrac{35}{50} =$

Lesson Summary

Any terminating decimal can be converted to a fraction using place value (e.g., 0.35 is thirty-five hundredths or $\dfrac{35}{100}$). A fraction whose denominator includes only factors of 2 and 5 can be converted to a decimal by writing the denominator as a power of ten.

Name _____   Date _____

1. Write 3.0035 as a fraction. Explain your process.

2. This week is just one of 40 weeks that you spend in the classroom this school year. Convert the fraction $\frac{1}{40}$ to decimal form.

EUREKA
MATH®

Lesson 13:   Converting Between Fractions and Decimals Using
              Equivalent Fractions

157

© 2019 Great Minds®. eureka-math.org

Converting Terminating Decimals to Fractions

Convert each terminating decimal to a fraction in its simplest form.

1. 0.8

> The decimal place farthest to the right is the tenths place. Therefore, my denominator will be 10, and my numerator will be 8.

$$0.8 = \frac{8}{10} = \frac{4}{5}$$

> The numerator and denominator have a greatest common factor of 2. I can divide this common factor out to write the fraction in simplest form.

2. 0.375

> The decimal place farthest to the right is the thousandths place. Therefore, my denominator will be 1,000, and my numerator will be 375.

$$0.375 = \frac{375}{1,000} = \frac{75}{200} = \frac{15}{40} = \frac{3}{8}$$

> Instead of finding the greatest common factor, I kept dividing by a smaller common factor until the only common factor between the numerator and denominator was 1.

3. 0.05

> The extra 0 does not change anything, I still look at the decimal place farthest to the right to determine the denominator.

$$0.05 = \frac{5}{100} = \frac{1}{20}$$

Converting Fractions to Decimals

I need to write each fraction with a denominator that is a power of 10.

Convert each fraction or mixed number to a decimal using an equivalent fraction.

In order to make my denominator a power of 10, I need to multiply both the numerator and denominator by 2.

4. $\frac{2}{5}$

$$\frac{2}{5} = \frac{2 \times 2}{5 \times 2} = \frac{4}{10} = 0.4$$

In order to make my denominator a power of 10, I need the same number of 2's and 5's in the denominator. Therefore, I need another factor of 5.

5. $\frac{7}{20}$

$$\frac{7}{20} = \frac{7}{2^2 \times 5} = \frac{7 \times 5}{2^2 \times 5^2} = \frac{35}{100} = 0.35$$

If the denominator is 100, then the decimal farthest to the right should be in the hundredths place.

6. $\frac{13}{250}$

This time I need two more factors of 2 to have the same number of 2's and 5's, which will make my denominator a power of 10.

$$\frac{13}{250} = \frac{13}{2 \times 5^3} = \frac{13 \times 2^2}{2^3 \times 5^3} = \frac{52}{1000} = 0.052$$

I need the decimal farthest to the right to be in the thousandths place, so I need to use a 0 for a place holder.

7. $\frac{21}{175}$

I have to divide out the factor of 7 from the numerator and denominator before I can change the denominator to a power of 10.

$$\frac{21}{175} = \frac{7 \times 3}{7 \times 5^2} = \frac{3}{5^2} = \frac{3 \times 2^2}{5^2 \times 2^2} = \frac{12}{100} = 0.12$$

Lesson 13: Converting Between Fractions and Decimals Using Equivalent Fractions

EUREKA MATH

1. Convert each terminating decimal to a fraction in its simplest form.
   a. 0.4
   b. 0.16
   c. 0.625
   d. 0.08
   e. 0.012

2. Convert each fraction or mixed number to a decimal using an equivalent fraction.
   a. $\dfrac{4}{5}$
   b. $\dfrac{3}{40}$
   c. $\dfrac{8}{200}$
   d. $3\dfrac{5}{16}$

3. Tanja is converting a fraction into a decimal by finding an equivalent fraction that has a power of 10 in the denominator. Sara looks at the last step in Tanja's work (shown below) and says that she cannot go any further. Is Sara correct? If she is, explain why. If Sara is incorrect, complete the remaining steps.

$$\frac{72}{480} = \frac{2^3 \cdot 3^2}{2^5 \cdot 3 \cdot 5}$$

Example 1: Can All Rational Numbers Be Written as Decimals?

a. Using the division button on your calculator, explore various quotients of integers 1 through 11. Record your fraction representations and their corresponding decimal representations in the space below.

b. What two types of decimals do you see?

Example 2: Decimal Representations of Rational Numbers

In the chart below, organize the fractions and their corresponding decimal representation listed in Example 1 according to their type of decimal.

What do these fractions have in common?   What do these fractions have in common?

Example 3: Converting Rational Numbers to Decimals Using Long Division

Use the long division algorithm to find the decimal value of $-\dfrac{3}{4}$.

Exercise 1

Convert each rational number to its decimal form using long division.

a. $-\dfrac{7}{8} =$

b. $\dfrac{3}{16} =$

EUREKA
MATH

Example 4: Converting Rational Numbers to Decimals Using Long Division

Use long division to find the decimal representation of $\frac{1}{3}$.

Exercise 2

Calculate the decimal values of the fraction below using long division. Express your answers using bars over the shortest sequence of repeating digits.

a. $-\frac{4}{9}$

b. $-\frac{1}{11}$

c. $\frac{1}{7}$

d. $-\frac{5}{6}$

Example 5: Fractions Represent Terminating or Repeating Decimals

How do we determine whether the decimal representation of a quotient of two integers, with the divisor not equal to zero, will terminate or repeat?

Example 6: Using Rational Number Conversions in Problem Solving

a. Eric and four of his friends are taking a trip across the New York State Thruway. They decide to split the cost of tolls equally. If the total cost of tolls is $8, how much will each person have to pay?

b. Just before leaving on the trip, two of Eric's friends have a family emergency and cannot go. What is each person's share of the $8 tolls now?

Lesson 14:   Converting Rational Numbers to Decimals Using
Long Division

© 2019 Great Minds®. eureka-math.org

Lesson Summary

The real world requires that we represent rational numbers in different ways depending on the context of a situation. All rational numbers can be represented as either terminating decimals or repeating decimals using the long division algorithm. We represent repeating decimals by placing a bar over the shortest sequence of repeating digits.

Name _____   Date _____

1. What is the decimal value of $\frac{4}{11}$?

2. How do you know that $\frac{4}{11}$ is a repeating decimal?

3. What causes a repeating decimal in the long division algorithm?

1. Convert each rational number into its decimal form.

   a. $\dfrac{3}{12}$

   $$\dfrac{3}{12} = \dfrac{3}{2^2 \times 3} = \dfrac{1}{2^2} = \dfrac{1 \times 5^2}{2^2 \times 5^2} = \dfrac{25}{100} = 0.25$$

   $$\dfrac{3}{12} = 0.25$$

   > I know the decimal will terminate because the fraction can be rewritten with a denominator that is a power of 10.

   > I notice the remainder continues to repeat, which means the digits in the quotient will also repeat.

   b. $\dfrac{1}{12}$

   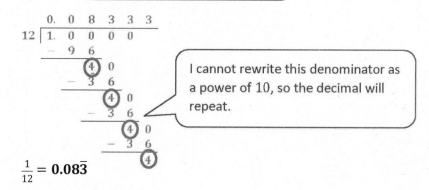

   > I cannot rewrite this denominator as a power of 10, so the decimal will repeat.

   $$\dfrac{1}{12} = 0.08\overline{3}$$

2. Josephine thinks $-\dfrac{5}{15}$ is a terminating decimal. Is Josephine correct? Why or why not?

   $$-\dfrac{5}{15} = -\dfrac{1}{3}$$

   Josephine is not correct because the denominator cannot be written as a power of 10, which must be true if the fraction represents a terminating decimal.

1. Convert each rational number into its decimal form.

$\dfrac{1}{9} =$ _____

$\dfrac{1}{6} =$ _____

$\dfrac{2}{9} =$ _____

$\dfrac{1}{3} =$ _____

$\dfrac{2}{6} =$ _____

$\dfrac{3}{9} =$ _____

$\dfrac{4}{9} =$ _____

$\dfrac{3}{6} =$ _____

$\dfrac{5}{9} =$ _____

$\dfrac{2}{3} =$ _____

$\dfrac{4}{6} =$ _____

$\dfrac{6}{9} =$ _____

$\dfrac{7}{9} =$ _____

$\dfrac{5}{6} =$ _____

$\dfrac{8}{9} =$ _____

One of these decimal representations is not like the others. Why?

Enrichment:

2. Chandler tells Aubrey that the decimal value of $-\frac{1}{17}$ is not a repeating decimal. Should Aubrey believe him? Explain.

3. Complete the quotients below without using a calculator, and answer the questions that follow.

   a. Convert each rational number in the table to its decimal equivalent.

$\frac{1}{11} =$	$\frac{2}{11} =$	$\frac{3}{11} =$	$\frac{4}{11} =$	$\frac{5}{11} =$
$\frac{6}{11} =$	$\frac{7}{11} =$	$\frac{8}{11} =$	$\frac{9}{11} =$	$\frac{10}{11} =$

   Do you see a pattern? Explain.

   b. Convert each rational number in the table to its decimal equivalent.

$\frac{0}{99} =$	$\frac{10}{99} =$	$\frac{20}{99} =$	$\frac{30}{99} =$	$\frac{45}{99} =$
$\frac{58}{99} =$	$\frac{62}{99} =$	$\frac{77}{99} =$	$\frac{81}{99} =$	$\frac{98}{99} =$

   Do you see a pattern? Explain.

   c. Can you find other rational numbers that follow similar patterns?

Converting Rational Numbers to Decimals Using Long Division

EUREKA MATH®

Number Correct: _____

Integer Multiplication—Round 1

Directions: Determine the product of the integers, and write it in the column to the right.

1.	$-2 \cdot -8$	
2.	$-4 \cdot 3$	
3.	$5 \cdot -7$	
4.	$1 \cdot -1$	
5.	$-6 \cdot 9$	
6.	$-2 \cdot -7$	
7.	$8 \cdot -3$	
8.	$0 \cdot -9$	
9.	$12 \cdot -5$	
10.	$-4 \cdot 2$	
11	$-1 \cdot -6$	
12.	$10 \cdot -4$	
13.	$14 \cdot -3$	
14.	$-5 \cdot -13$	
15.	$-16 \cdot -8$	
16.	$18 \cdot -2$	
17.	$-15 \cdot 7$	
18.	$-19 \cdot 1$	
19.	$12 \cdot 12$	
20.	$9 \cdot -17$	
21.	$-8 \cdot -14$	
22.	$-7 \cdot 13$	

23.	$-14 \cdot -12$	
24.	$15 \cdot -13$	
25.	$16 \cdot -18$	
26.	$24 \cdot -17$	
27.	$-32 \cdot -21$	
28.	$19 \cdot -27$	
29.	$-39 \cdot 10$	
30.	$43 \cdot 22$	
31.	$11 \cdot -33$	
32.	$-29 \cdot -45$	
33.	$37 \cdot -44$	
34.	$-87 \cdot -100$	
35.	$92 \cdot -232$	
36.	$456 \cdot 87$	
37.	$-143 \cdot 76$	
38.	$439 \cdot -871$	
39.	$-286 \cdot -412$	
40.	$-971 \cdot 342$	
41.	$-773 \cdot -407$	
42.	$-820 \cdot 638$	
43.	$591 \cdot -734$	
44.	$491 \cdot -197$	

EUREKA MATH®

Integer Multiplication—Round 2

Number Correct: _____

Improvement: _____

Directions: Determine the product of the integers, and write it in the column to the right.

1.	$-9 \cdot -7$		23.	$-22 \cdot 14$	
2.	$0 \cdot -4$		24.	$-18 \cdot -32$	
3.	$3 \cdot -5$		25.	$-24 \cdot 19$	
4.	$6 \cdot -8$		26.	$47 \cdot 21$	
5.	$-2 \cdot 1$		27.	$17 \cdot -39$	
6.	$-6 \cdot 5$		28.	$-16 \cdot -28$	
7.	$-10 \cdot -12$		29.	$-67 \cdot -81$	
8.	$11 \cdot -4$		30.	$-36 \cdot 44$	
9.	$3 \cdot 8$		31.	$-50 \cdot 23$	
10.	$12 \cdot -7$		32.	$66 \cdot -71$	
11.	$-1 \cdot 8$		33.	$82 \cdot -29$	
12.	$5 \cdot -10$		34.	$-32 \cdot 231$	
13.	$3 \cdot -13$		35.	$89 \cdot -744$	
14.	$15 \cdot -8$		36.	$623 \cdot -22$	
15.	$-9 \cdot 14$		37.	$-870 \cdot -46$	
16.	$-17 \cdot 5$		38.	$179 \cdot 329$	
17.	$16 \cdot 2$		39.	$-956 \cdot 723$	
18.	$19 \cdot -7$		40.	$874 \cdot -333$	
19.	$-6 \cdot 13$		41.	$908 \cdot -471$	
20.	$1 \cdot -18$		42.	$-661 \cdot -403$	
21.	$-14 \cdot -3$		43.	$-520 \cdot -614$	
22.	$-10 \cdot -17$		44.	$-309 \cdot 911$	

Exercise 1

a. In the space below, create a word problem that involves integer multiplication. Write an equation to model the situation.

b. Now change the word problem by replacing the integers with non-integer rational numbers (fractions or decimals), and write the new equation.

c. Was the process used to solve the second problem different from the process used to solve the first? Explain.

d. The Rules for Multiplying Rational Numbers are the same as the Rules for Multiplying Integers:

1. _____

2. _____

3. _____

EUREKA
MATH

Exercise 2

a.   In one year, Melinda's parents spend $2,640.90 on cable and internet service. If they spend the same amount each month, what is the resulting monthly change in the family's income?

b.   The Rules for Dividing Rational Numbers are the same as the Rules for Dividing Integers:

1.   _____

2.   _____

3.   _____

Exercise 3

Use the fundraiser chart to help answer the questions that follow.

Grimes Middle School Flower Fundraiser

Customer	Plant Type	Number of Plants	Price per Plant	Total	Paid? Yes or No
Tamara Jones	tulip	2	$4.25		No
Mrs. Wolff	daisy	1	$3.75	$3.75	Yes
Mr. Clark	geranium	5	$2.25		Yes
Susie (Jeremy's sister)	violet	1	$2.50	$2.50	Yes
Nana and Pop (Jeremy's grandparents)	daisy	4	$3.75	$15.00	No

Jeremy is selling plants for the school's fundraiser, and listed above is a chart from his fundraiser order form. Use the information in the chart to answer the following questions. Show your work, and represent the answer as a rational number; then, explain your answer in the context of the situation.

a. If Tamara Jones writes a check to pay for the plants, what is the resulting change in her checking account balance?

Numerical Answer:

Explanation:

b. Mr. Clark wants to pay for his order with a $20 bill, but Jeremy does not have change. Jeremy tells Mr. Clark he will give him the change later. How will this affect the total amount of money Jeremy collects? Explain. What rational number represents the change that must be made to the money Jeremy collects?

Numerical Answer:

Explanation:

c. Jeremy's sister, Susie, borrowed the money from their mom to pay for her order. Their mother has agreed to deduct an equal amount of money from Susie's allowance each week for the next five weeks to repay the loan. What is the weekly change in Susie's allowance?

Numerical Answer:

Explanation:

d. Jeremy's grandparents want to change their order. They want to order three daisies and one geranium, instead of four daisies. How does this change affect the amount of their order? Explain how you arrived at your answer.

e. Jeremy approaches three people who do not want to buy any plants; however, they wish to donate some money for the fundraiser when Jeremy delivers the plants one week later. If the people promise to donate a total of $14.40, what will be the average cash donation?

f. Jeremy spends one week collecting orders. If 22 people purchase plants totaling $270, what is the average amount of Jeremy's sale?

Lesson 15:   Multiplication and Division of Rational Numbers

Lesson Summary

The rules that apply for multiplying and dividing integers apply to rational numbers. We can use the products and quotients of rational numbers to describe real-world situations.

Name _____   Date _____

Harrison made up a game for his math project. It is similar to the Integer Game; however, in addition to integers, there are cards that contain other rational numbers such as −0.5 and −0.25. Write a multiplication or division equation to represent each problem below. Show all related work.

1. Harrison discards three −0.25 cards from his hand. How does this affect the overall point value of his hand? Write an equation to model this situation.

2. Ezra and Benji are playing the game with Harrison. After Ezra doubles his hand's value, he has a total of −14.5 points. What was his hand's value before he doubled it?

3. Benji has four −0.5 cards. What is his total score?

1. Charlotte owes her parents $135. If Charlotte pays her parents $15 every week for 7 weeks, how much money will she still owe her parents?

> To determine how much Charlotte pays her parents, I can multiply the size of the payment by the number of payments.

$$135 + 7(-15) = 135 + (-105) = 30$$

Charlotte will still owe her parents $30 *after making 7 equal payments of* $15.

2. Find at least two sets of values that will make each equation true.

    a. Fill in the blanks with two rational numbers that will make the equation true.

$$\underline{\quad} \times \left(-\frac{1}{5}\right) \times \underline{\quad} = 10$$

What must be true about the relationship between the two numbers you choose?

Two possible answers: -10 *and* 5 *or* 10 *and* -5

The two numbers must be factors of 50 *and have opposite signs.*

> I can use my knowledge of solving equations to determine the product of the two missing values.

> To get a positive quotient, I need an even number of negative factors. The factor $\left(-\frac{1}{5}\right)$ is already negative, so one of the other two factors needs to be negative.

> This part of the equation has a value of 5, which means the missing two factors should have a product of −12.

b. Fill in the blanks with two rational numbers that will make the equation true.

$$(-2.5) \times 50 \div (-25) \times \underline{\quad} \times \underline{\quad} = -60$$

What must be true about the relationship between the two numbers you choose?

Two possible answers: 2 and −6 or −3 and 4

The two numbers must be factors of 12 and have opposite signs.

> To have a negative answer, I must have an odd number of negative factors. Two of the factors are already negative, which means one additional factor needs to be negative.

3. Create a word problem that can be represented by the expression, and then represent the quotient as a single rational number.

$$-10 \div 2\frac{1}{2}$$

The temperature dropped 10 degrees in $2\frac{1}{2}$ hours. If the temperature dropped at a constant rate, how much did the temperature drop each hour?

$$-10 \div 2\frac{1}{2}$$

$$-10 \div \frac{5}{2}$$

$$-\frac{10}{1} \times \frac{2}{5}$$

$$-\frac{20}{5}$$

$$-4$$

> I could use a different real-world problem, but the answer is still −4. Only the unit will change!

The temperature dropped 4 degrees each hour.

1. At lunch time, Benjamin often borrows money from his friends to buy snacks in the school cafeteria. Benjamin borrowed $0.75 from his friend Clyde five days last week to buy ice cream bars. Represent the amount Benjamin borrowed as the product of two rational numbers; then, determine how much Benjamin owed his friend last week.

2. Monica regularly records her favorite television show. Each episode of the show requires 3.5% of the total capacity of her video recorder. Her recorder currently has 62% of its total memory free. If Monica records all five episodes this week, how much space will be left on her video recorder?

For Problems 3–5, find at least two possible sets of values that will work for each problem.

3. Fill in the blanks with two rational numbers (other than 1 and –1). ____ × $\left(-\frac{1}{2}\right)$ × ____ = –20
   What must be true about the relationship between the two numbers you chose?

4. Fill in the blanks with two rational numbers (other than 1 and –1). $-5.6 \times 100 \div 80 \times$ ____ × ____ = 700
   What must be true about the relationship between the two numbers you chose?

5. Fill in the blanks with two rational numbers. ____ × ____ = –0.75
   What must be true about the relationship between the two numbers you chose?

For Problems 6–8, create word problems that can be represented by each expression, and then represent each product or quotient as a single rational number.

6. $8 \times (-0.25)$

7. $-6 \div \left(1\frac{1}{3}\right)$

8. $-\frac{1}{2} \times 12$

Number Correct: _____

Integer Division—Round 1

Directions: Determine the quotient of the integers, and write it in the column to the right.

1.	$4 \div 1$		23.	$-16 \div (-4)$	
2.	$4 \div (-1)$		24.	$16 \div (-2)$	
3.	$-4 \div (-1)$		25.	$-16 \div 4$	
4.	$-4 \div 1$		26.	$-20 \div 4$	
5.	$6 \div 2$		27.	$-20 \div (-4)$	
6.	$-6 \div (-2)$		28.	$-28 \div 4$	
7.	$-6 \div 2$		29.	$28 \div (-7)$	
8.	$6 \div -2$		30.	$-28 \div (-7)$	
9.	$8 \div (-4)$		31.	$-40 \div (-5)$	
10.	$-8 \div (-4)$		32.	$56 \div (-7)$	
11.	$-8 \div 4$		33.	$96 \div (-3)$	
12.	$8 \div 4$		34.	$-121 \div (-11)$	
13.	$9 \div (-3)$		35.	$169 \div (-13)$	
14.	$-9 \div 3$		36.	$-175 \div (25)$	
15.	$-10 \div 5$		37.	$1 \div 4$	
16.	$10 \div (-2)$		38.	$-1 \div 4$	
17.	$-10 \div (-2)$		39.	$-1 \div (-4)$	
18.	$-10 \div (-5)$		40.	$-3 \div (-4)$	
19.	$-14 \div 7$		41.	$-5 \div 20$	
20.	$14 \div (-2)$		42.	$6 \div (-18)$	
21.	$-14 \div (-2)$		43.	$-24 \div 48$	
22.	$-14 \div (-7)$		44.	$-16 \div 64$	

Number Correct: _____

Improvement: _____

Integer Division—Round 2

Directions: Determine the quotient of the integers, and write it in the column to the right.

1.	$5 \div 1$		23.	$-18 \div (-9)$	
2.	$5 \div (-1)$		24.	$18 \div (-2)$	
3.	$-5 \div (-1)$		25.	$-18 \div 9$	
4.	$-5 \div 1$		26.	$-24 \div 4$	
5.	$6 \div 3$		27.	$-24 \div (-4)$	
6.	$-6 \div (-3)$		28.	$-24 \div 6$	
7.	$-6 \div 3$		29.	$30 \div (-6)$	
8.	$6 \div -3$		30.	$-30 \div (-5)$	
9.	$8 \div (-2)$		31.	$-48 \div (-6)$	
10.	$-8 \div (-2)$		32.	$64 \div (-4)$	
11.	$-8 \div 2$		33.	$105 \div (-7)$	
12.	$8 \div 2$		34.	$-144 \div (-12)$	
13.	$-9 \div (-3)$		35.	$196 \div (-14)$	
14.	$9 \div 3$		36.	$-225 \div 25$	
15.	$-12 \div 6$		37.	$2 \div 4$	
16.	$12 \div (-2)$		38.	$-2 \div 4$	
17.	$-12 \div (-2)$		39.	$-2 \div (-4)$	
18.	$-12 \div (-6)$		40.	$-4 \div (-8)$	
19.	$-16 \div 8$		41.	$-5 \div 40$	
20.	$16 \div (-2)$		42.	$6 \div (-42)$	
21.	$-16 \div (-2)$		43.	$-25 \div 75$	
22.	$-16 \div (-8)$		44.	$-18 \div 108$	

Example 1: Using the Commutative and Associative Properties to Efficiently Multiply Rational Numbers

a. Evaluate the expression below.

$$-6 \times 2 \times (-2) \times (-5) \times (-3)$$

b. What types of strategies were used to evaluate the expressions?

c. Can you identify the benefits of choosing one strategy versus another?

d. What is the sign of the product, and how was the sign determined?

Exercise 1

Find an efficient strategy to evaluate the expression and complete the necessary work.

$$-1 \times (-3) \times 10 \times (-2) \times 2$$

Exercise 2

Find an efficient strategy to evaluate the expression and complete the necessary work.

$$4 \times \frac{1}{3} \times (-8) \times 9 \times \left(-\frac{1}{2}\right)$$

Exercise 3

What terms did you combine first and why?

Lesson 16:   Applying the Properties of Operations to Multiply and
                        Divide Rational Numbers

EUREKA
MATH®

Exercise 4

Refer to the example and exercises. Do you see an easy way to determine the sign of the product first?

Example 2: Using the Distributive Property to Multiply Rational Numbers

Rewrite the mixed number as a sum; then, multiply using the distributive property.

$$-6 \times \left(5\frac{1}{3}\right)$$

Exercise 5

Multiply the expression using the distributive property.

$$9 \times \left(-3\frac{1}{2}\right)$$

EUREKA
MATH®

© 2019 Great Minds®. eureka-math.org

Example 3: Using the Distributive Property to Multiply Rational Numbers

Evaluate using the distributive property.

$$16 \times \left(-\frac{3}{8}\right) + 16 \times \frac{1}{4}$$

Example 4: Using the Multiplicative Inverse to Rewrite Division as Multiplication

Rewrite the expression as only multiplication and evaluate.

$$1 \div \frac{2}{3} \times (-8) \times 3 \div \left(-\frac{1}{2}\right)$$

Exercise 6

$$4.2 \times \left(-\frac{1}{3}\right) \div \frac{1}{6} \times (-10)$$

Lesson 16:   Applying the Properties of Operations to Multiply and
Divide Rational Numbers

EUREKA
MATH®

Lesson Summary

Multiplying and dividing using the strict order of the operations in an expression is not always efficient. The properties of multiplication allow us to manipulate the expression by rearranging and regrouping factors that are easier to compute (like grouping factors 2 and 5 to get 10).

Where division is involved, we can easily rewrite the division by a number as multiplication by its reciprocal, and then use the properties of multiplication.

If an expression is only a product of factors, then the sign of its value is easily determined by the number of negative factors: the sign is positive if there are an even number of negative factors and negative if there is an odd number of factors.

Name _____   Date _____

1. Evaluate the expression below using the properties of operations.

$$18 \div \left(-\frac{2}{3}\right) \times 4 \div (-7) \times (-3) \div \left(\frac{1}{4}\right)$$

2.
   a. Given the expression below, what will the sign of the product be? Justify your answer.

$$-4 \times \left(-\frac{8}{9}\right) \times 2.78 \times \left(1\frac{1}{3}\right) \times \left(-\frac{2}{5}\right) \times (-6.2) \times (-0.2873) \times \left(3\frac{1}{11}\right) \times A$$

   b. Give a value for A that would result in a positive value for the expression.

   c. Give a value for A that would result in a negative value for the expression.

EUREKA MATH®

1. Evaluate the expression $\left(-\frac{1}{5}\right) \times (-8) \div \left(-\frac{1}{3}\right) \times 15$

   a. Using order of operations only.

   > Using the order of operations to evaluate this expression, I complete the operations from left to right.

   $$\left(-\frac{1}{5}\right) \times (-8) \div \left(-\frac{1}{3}\right) \times 15$$

   $$\frac{8}{5} \div \left(-\frac{1}{3}\right) \times 15$$

   $$\frac{8}{5} \times (-3) \times 15$$

   $$\left(-\frac{24}{5}\right) \times 15$$

   $$-72$$

   b. Using the properties and methods used in Lesson 16.

   > I use the commutative property to change the order of the factors. This allows me to eliminate the fractions.

   $$\left(-\frac{1}{5}\right) \times (-8) \div \left(-\frac{1}{3}\right) \times 15$$

   $$\left(-\frac{1}{5}\right) \times 15 \times (-8) \times (-3)$$

   $$(-3) \times (-8) \times (-3)$$

   $$-72$$

2. Evaluate each expression using the distributive property.

   a. $3\frac{1}{3} \times (-12)$

   $$\left(3 + \frac{1}{3}\right) \times (-12)$$

   > $3\frac{1}{3}$ is equivalent to $3 + \frac{1}{3}$.

   $$(3) \times (-12) + \frac{1}{3} \times (-12)$$

   > I distribute the -12 to both values in the parentheses.

   $$-36 + (-4)$$

   $$-40$$

b. $\frac{3}{4}(-3) + \frac{3}{4}(11)$

$\frac{3}{4}(-3 + 11)$

I use the distributive property to factor out the common factor $\left(\frac{3}{4}\right)$. This allows me to combine the integers before multiplying by the fractional value.

$\frac{3}{4}(8)$

6

3. Examine the problem and work below. Find and explain the errors, and then find the correct value of the expression.

$(-3) \times 0.4 \times 2 \div \left(-\frac{1}{5}\right) \div 3$

$(-3) \times 0.4 \times 2 \times (-5) \times 3$

When division is changed to multiplication, the divisor needs to be inverted.

$(-3) \times 0.4 \times (-5) \times 2 \times 3$

$(-3) \times (-2) \times 2 \times 3$

-36

Two negative factors result in a positive product.

There are two mistakes in the work. First, when changing $\div 3$ to multiplication, it should be changed to $\times \frac{1}{3}$ because we need to invert and multiply. Second, there is an error with the negative signs in the final step. An even number of negative factors will result in a positive product, not a negative product. The correct work is shown below.

$(-3) \times 0.4 \times 2 \times (-5) \times \frac{1}{3}$

$(-3) \times \frac{1}{3} \times 0.4 \times (-5) \times 2$

I can use the commutative property like I did with an earlier problem.

$-1 \times 0.4 \times (-5) \times 2$

$-1 \times (-2) \times 2$

4

EUREKA MATH

1. Evaluate the expression $-2.2 \times (-2) \div \left(-\frac{1}{4}\right) \times 5$

    a. Using the order of operations only.

    b. Using the properties and methods used in Lesson 16.

    c. If you were asked to evaluate another expression, which method would you use, (a) or (b), and why?

2. Evaluate the expressions using the distributive property.

    a. $\left(2\frac{1}{4}\right) \times (-8)$

    b. $\frac{2}{3}(-7) + \frac{2}{3}(-5)$

3. Mia evaluated the expression below but got an incorrect answer. Find Mia's error(s), find the correct value of the expression, and explain how Mia could have avoided her error(s).

    $0.38 \times 3 \div \left(-\frac{1}{20}\right) \times 5 \div (-8)$

    $0.38 \times 5 \times \left(\frac{1}{20}\right) \times 3 \times (-8)$

    $0.38 \times \left(\frac{1}{4}\right) \times 3 \times (-8)$

    $0.38 \times \left(\frac{1}{4}\right) \times (-24)$

    $0.38 \times (-6)$

    -2.28

Opening Exercise

For his birthday, Zack and three of his friends went to a movie. They each got a ticket for $8.00 and the same snack from the concession stand. If Zack's mom paid $48 for the group's tickets and snacks, how much did each snack cost?

The equation $4(s + 8) = 48$ represents the situation when s represents the cost, in dollars, of one snack.

Exploratory Challenge: Expenses on Your Family Vacation

John and Ag are summarizing some of the expenses of their family vacation for themselves and their three children, Louie, Missy, and Bonnie. Write an algebraic equation, create a model to determine how much each item will cost using all of the given information, and answer the questions that follow.

Expenses:

Car and insurance fees: $400	Airfare and insurance fees: $875	Motel and tax: $400
Baseball game and hats: $103.83	Movies for one day: $75	Soda and pizza: $37.95
	Sandals and T-shirts: $120	

Your Group's Scenario Solution:

After collaborating with all of the groups, summarize the findings in the table below.

Cost of Evening Movie	
Cost of 1 Slice of Pizza	
Cost of the Admission Ticket to the Baseball Game	
Cost of 1 T-Shirt	
Cost of 1 Airplane Ticket	
Daily Cost for Car Rental	
Nightly Charge for Motel	

Using the results, determine the cost of the following:

1. A slice of pizza, 1 plane ticket, 2 nights in the motel, and 1 evening movie.

2. One T-shirt, 1 ticket to the baseball game, and 1 day of the rental car.

Exercise

The cost of a babysitting service on a cruise is $10 for the first hour and $12 for each additional hour. If the total cost of babysitting baby Aaron was $58, how many hours was Aaron at the sitter?

Exploratory Challenge Scenarios

Scenario 1

During one rainy day on the vacation, the entire family decided to go watch a matinee movie in the morning and a drive-in movie in the evening. The price for a matinee movie in the morning is different than the cost of a drive-in movie in the evening. The tickets for the matinee morning movie cost $6 each. How much did each person spend that day on movie tickets if the ticket cost for each family member was the same? What was the cost for a ticket for the drive-in movie in the evening?

Scenario 2

For dinner one night, the family went to the local pizza parlor. The cost of a soda was $3. If each member of the family had a soda and one slice of pizza, how much did one slice of pizza cost?

Scenario 3

One night, John, Louie, and Bonnie went to see the local baseball team play a game. They each bought a game ticket and a hat that cost $10. How much was each ticket to enter the ballpark?

Scenario 4

While John, Louie, and Bonnie went to see the baseball game, Ag and Missy went shopping. They bought a T-shirt for each member of the family and bought two pairs of sandals that cost $10 a pair. How much was each T-shirt?

Scenario 5

The family flew in an airplane to their vacation destination. Each person had to have his own ticket for the plane and also pay $25 in insurance fees per person. What was the cost of one ticket?

Scenario 6

While on vacation, the family rented a car to get them to all the places they wanted to see for five days. The car costs a certain amount each day, plus a one-time insurance fee of $50. How much was the daily cost of the car (not including the insurance fees)?

Scenario 7

The family decided to stay in a motel for four nights. The motel charges a nightly fee plus $60 in state taxes. What is the nightly charge with no taxes included?

Lesson Summary

Tape diagrams can be used to model and identify the sequence of operations to find a solution algebraically.

The goal in solving equations algebraically is to isolate the variable.

The process of doing this requires *undoing* addition or subtraction to obtain a 0 and *undoing* multiplication or division to obtain a 1. The additive inverse and multiplicative inverse properties are applied to get the 0 (the additive identity) and 1 (the multiplicative identity).

The addition and multiplication properties of equality are applied because in an equation, $A = B$, when a number is added or multiplied to both sides, the resulting sum or product remains equal.

EUREKA
MATH

Name _____   Date _____

1. Eric's father works two part-time jobs, one in the morning and one in the afternoon, and works a total of 40 hours each 5-day workweek. If his schedule is the same each day, and he works 3 hours each morning, how many hours does Eric's father work each afternoon?

2. Henry is using a total of 16 ft. of lumber to make a bookcase. The left and right sides of the bookcase are each 4 ft. high. The top, bottom, and two shelves are all the same length, labeled S. How long is each shelf?

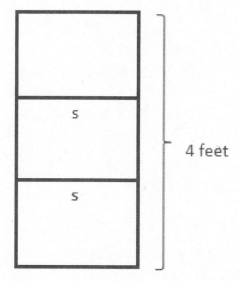

Solve each problem by writing an equation and constructing a tape diagram.

1. Liam always eats 2 servings of fruit every day. He also eats some vegetables every day. He eats 35 servings of fruits and vegetables every week. If Liam eats the same number of vegetables every day, how many servings of vegetables does he eat each day?

Algebraic Equation

Let v represent the number of servings of vegetables Liam eats each day.

I add the number of servings of fruits and vegetables together and then multiply by 7 because there are 7 days in a week, and I know how many servings Liam eats in a week, 35.

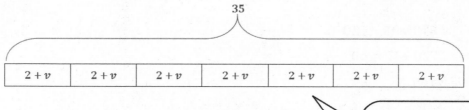

$$7(2+v) = 35$$
$$14 + 7v = 35$$
$$14 - 14 + 7v = 35 - 14$$
$$7v = 21$$
$$\left(\frac{1}{7}\right)(7v) = \left(\frac{1}{7}\right)(21)$$
$$v = 3$$

Here, I apply the distributive property. I could multiply by the multiplicative inverse instead.

I can collect like terms and then multiply by the multiplicative inverse to determine the value of the variable.

Liam eats 3 servings of vegetables each day.

Tape Diagram

35

$2+v$	$2+v$	$2+v$	$2+v$	$2+v$	$2+v$	$2+v$

$7v$

$7(2) = 14$

$35 - 14 = 21$

$21 \div 7 = 3$

I have 7 equal sections, one for each day of the week. Each section shows the number of servings of fruits and vegetables Liam eats each day.

Liam eats 3 servings ofvegetobles each day.

2. Ava bought her first cell phone for $95 and now has a monthly bill. If Ava paid $815 during the first year of having a cell phone, what is the amount of Ava's monthly bill?

Algebraic Equation

Let m represent the amount of Ava's monthly bill.

$$12m + 95 = 815$$
$$12m + 95 - 95 = 815 - 95$$
$$12m = 720$$
$$\left(\frac{1}{12}\right)(12m) = \left(\frac{1}{12}\right)(720)$$
$$m = 60$$

> I know Ava makes 12 equal monthly payments over the course of the year. The sum of her monthly bill and her cell phone will be the amount she paid during the first year.

Ava's monthly bill is $60.

> I create a tape diagram that shows 12 equal payments of m plus the cost of the cell phone, $95. The total length of the tape diagram represents the amount Ava paid during the first year, $815.

Tape Diagram

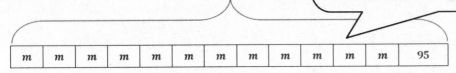

815

| m | m | m | m | m | m | m | m | m | m | m | m | 95 |

$12m$ $815 - 95 = 720$

$720 \div 12 = 60$

Ava's monthly bill is $60.

EUREKA MATH

1. A taxi cab in Myrtle Beach charges $2 per mile and $1 for every person. If a taxi cab ride for two people costs $12, how far did the taxi cab travel?

   5 miles

2. Heather works as a waitress at her family's restaurant. She works 2 hours every morning during the breakfast shift and returns to work each evening for the dinner shift. In the last four days, she worked 28 hours. If Heather works the same number of hours every evening, how many hours did she work during each dinner shift?

3. Julian exercises 5 times a week. She runs 3 miles each morning and bikes in the evening. If she exercises a total of 30 miles for the week, how many miles does she bike each evening?

4. Marc eats an egg sandwich for breakfast and a big burger for lunch every day. The egg sandwich has 250 calories. If Marc has 5,250 calories for breakfast and lunch for the week in total, how many calories are in one big burger?

5. Jackie won tickets playing the bowling game at the local arcade. The first time, she won 60 tickets. The second time, she won a bonus, which was 4 times the number of tickets of the original second prize. Altogether she won 200 tickets. How many tickets was the original second prize?

1. $px + a = r$

   #2 per mile   #1 per person
      #12 for 2 people

   $2m + 2 = 12$ $\dfrac{2m}{2} = \dfrac{10}{2}$
   $\quad -2 \ -2$
   $\overline{\quad 0 \quad 10}$ $m = 5$

Exercise 1

John's father asked him to compare several different cell phone plans and identify which plan will be the least expensive for the family. Each phone company charges a monthly fee, but this fee does not cover any services: phone lines, texting, or internet access. Use the information contained in the table below to answer the following questions.

Cell Phone Plans

Name of Plan	Monthly Fee (Includes 1,500 shared minutes)	Price per Phone Line x	Price per line for Unlimited Texting y	Price per line for Internet Access z
Company A	$70	$20	$15	$15
Company B	$90	$15	$10	$20
Company C	$200	$10	included in monthly fee	included in monthly fee

All members of the family may not want identical plans; therefore, we will let x represent the number of phone lines, y represent the number of phone lines with unlimited texting, and z represent the number of phone lines with internet access.

Expression

Company A _____

Company B _____

Company C _____

Using the expressions above, find the cost to the family of each company's phone plan if:

a. Four people want a phone line, four people want unlimited texting, and the family needs two internet lines.

Company A	Company B	Company C

Which cell phone company should John's family use? Why?

b. Four people want a phone line, four people want unlimited texting, and all four people want internet lines.

Company A	Company B	Company C

Which cell phone company should John's family use? Why?

Lesson 18: Writing, Evaluating, and Finding Equivalent Expressions with Rational Numbers

EUREKA MATH

c.   Two people want a phone line, two people want unlimited texting, and the family needs two internet lines.

Company A	Company B	Company C

Which cell phone company should John's family use? Why?

Exercise 2

Three friends went to the movies. Each purchased a medium-sized popcorn for p dollars and a small soft drink for s dollars.

a.   Write the expression that represents the total amount of money (in dollars) the three friends spent at the concession stand.

b.   If the concession stand charges $6.50 for a medium-sized popcorn and $4.00 for a small soft drink, how much did the three friends spend on their refreshments altogether?

EUREKA
MATH®

Lesson 18:   Writing, Evaluating, and Finding Equivalent Expressions
                with Rational Numbers

221

© 2019 Great Minds®. eureka-math.org

Exercise 3

Complete the table below by writing equivalent expressions to the given expression and evaluating each expression with the given values.

Equivalent Expressions			
EXAMPLE: Evaluate $x = 2,$ $y = -1$	$4(x + 2y)$ $4(2 + 2(-1))$ $4(0)$ 0	$4x + 8y$ $4(2) + 8(-1)$ $8 + (-8)$ 0	$4x + 4y + 4y$ $4(2) + 4(-1) + 4(-1)$ $8 + (-4) + (-4)$ 0
1. Evaluate $y = 1$	$5(3 - 4y)$		
2. Evaluate $x = 5,$ $y = -2$	$-3x + 12y$		

Lesson 18: Writing, Evaluating, and Finding Equivalent Expressions
with Rational Numbers

EUREKA
MATH

© 2019 Great Minds®. eureka-math.org

3. Evaluate $x = -\dfrac{1}{2},$ $y = 1$			$-2x + 10x - 6y$

Lesson Summary

- An expression is a number or a letter, which can be raised to a whole number exponent. An expression can be a product whose factors are any one of the entities described above. An expression can also be the sum or difference of the products described above.

- To evaluate an expression, replace each variable with its corresponding numerical value. Using order of operations, the expression can be written as a single numerical value.

- When numbers are substituted into all the letters in an expression and the results are the same, then the expressions are equivalent.

EUREKA
MATH

Name _____   Date _____

Bradley and Louie are roommates at college. At the beginning of the semester, they each paid a security deposit of A dollars. When they move out, their landlord will deduct from this deposit any expenses (B) for excessive wear and tear and refund the remaining amount. Bradley and Louie will share the expenses equally.

- ▪ Write an expression that describes the amount each roommate will receive from the landlord when the lease expires.

- ▪ Evaluate the expression using the following information: Each roommate paid a $125 deposit, and the landlord deducted $50 total for damages.

1. Geraldine receives a weekly allowance. Every week she spends $2 on snacks at lunch time and saves the rest of her money.

   a. Write an expression that represents the amount Geraldine will save in 8 weeks if she receives a dollars each week for her allowance.

      Let a represent Geraldine's weekly allowance, in dollars.

      $$8(a-2)$$

      $$8a-16$$

      > I know that Geraldine saves her allowance minus the $2 she spends on snacks for 8 weeks.

      > I can apply the distributive property to write an equivalent expression.

   b. If Geraldine receives $10 each week for her allowance, how much money will she save in 8 weeks?

$8(a-2)$	**OR**	$8a-16$
$8(10-2)$		$8(10)-16$
$8(8)$		$80-16$
64		64

      Geraldine would save $64.

      > I now know the value of a and can substitute this value into either of the equivalent expressions.

2. During Nero's last basketball game, he made 6 field goals, 2 three-pointers, and 5 free throws.

   a. Write an expression to represent the total points Nero scored during the game.

      Let f represent the number of points for a field goal, p represent the number of points for a three-pointer, and t represent the number of points for a free throw.

      $$6f + 2p + 5t$$

      > I multiply the number of each type of shot Nero made by the number of points earned for each shot.

b. Write another expression that is equivalent to the one written above.

$3f + 3f + 2p + 5t$

> There are many options for an equivalent expression. I could have just changed the order of the terms.

c. If each field goal is worth 2 points, each three-pointer is worth 3 points, and each free throw is worth 1 point, how many total points did Nero score?

$6f + 2p + 5t$

$6(2) + 2(3) + 5(1)$

> I substitute each value in for the corresponding variable and use order of operations to evaluate the expression.

$12 + 6 + 5$

23

Nero scored 23 points.

3. The seventh grade student council is completing a fundraiser at a track meet. They are selling water bottles for $1.75 but paid $0.50 for each bottle of water. In order to keep the water cold, the student council also purchased a large cooler for $75. The table below shows the earnings, expenses, and profit earned when 80, 90, and 100 water bottles are sold.

Amount of Water Bottles Sold	Earnings (in dollars)	Expenses (in dollars)	Profit (in dollars)
80	$80(1.75) = 140$	$80(0.50) + 75 = 115$	$140 - 115 = 25$
90	$90(1.75) = 157.5$	$90(0.50) + 75 = 120$	$157.5 - 120 = 37.5$
100	$100(1.75) = 175$	$100(0.50) + 75 = 125$	$175 - 125 = 50$

a. Write an expression that represents the profit (in dollars) the student council earned by selling water bottles at the track meet.

Let w represent the number of water bottles sold.

$1.75w - 0.5w - 75$

$1.25w - 75$

> The first term shows the earnings. The last two terms are the expenses, so they must be subtracted from the earnings.

> I can collect like terms to write an equivalent expression.

Lesson 18: Writing, Evaluating, and Finding Equivalent Expressions with Rational Numbers

EUREKA MATH

b. How much profit did the student council make if it sold 50 water bottles? What does this mean?
    Explain why this might be the case.

$$1.25w - 75$$

$$1.25(50) - 75$$

$$62.5 - 75$$

$$-12.5$$

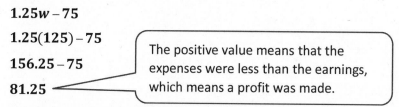

The negative value means that the expenses were more than the earnings, which means no profit was made.

The student council did not make any money; in fact, they lost $12.50. Possible reasons could be that it was not too hot and most people brought their own water to the track meet.

c. How much profit did the student council make if it sold 125 water bottles? What does this mean?
    Explain why this might be the case.

$$1.25w - 75$$

$$1.25(125) - 75$$

$$156.25 - 75$$

$$81.25$$

The positive value means that the expenses were less than the earnings, which means a profit was made.

The student council would make a profit of $81.25. The high number of water bottles sold could be explained by extremely hot weather or that most people did not bring enough water to the track meet.

EUREKA
MATH®

Lesson 18:   Writing, Evaluating, and Finding Equivalent Expressions
             with Rational Numbers

© 2019 Great Minds®. eureka-math.org

229

Example 1: Tic-Tac-Toe Review

Fill in the 9 spaces with one expression from the list below. Use one expression per space. You will use 9 of the expressions:

$12 - 4x$

$8x + 4 - 12x$

$8\left(\dfrac{1}{2}x - 2\right)$

$12 - 6x + 2x$

$-4x + 4$

$x - 2 + 2x - 4$

$4x - 12$

$4(x - 4)$

$3(x - 2)$

$0.1(40x) - \dfrac{1}{2}(24)$

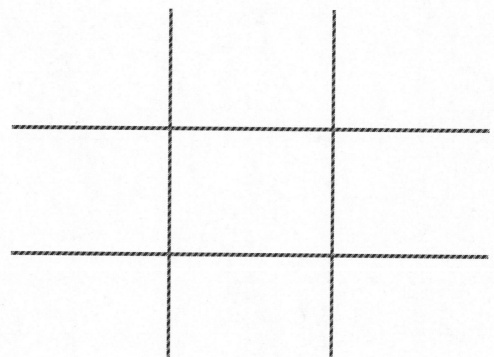

Example 2

Original Price (100%)	Discount Amount (20% Off)	New Price (Pay 80%)	Expression
100			
50			
28			
14.50			
x			

Lesson 19:   Writing, Evaluating, and Finding Equivalent Expressions with Rational Numbers

Example 3

An item that has an original price of x dollars is discounted 33%.

    a.   Write an expression that represents the amount of the discount.

    b.   Write two equivalent expressions that represent the new, discounted price.

    c.   Use one of your expressions to calculate the new, discounted price if the original price was $56.

    d.   How would the expressions you created in parts (a) and (b) have to change if the item's price had increased by 33% instead of decreased by 33%?

Example 4

Original Price (100%)	Discount (20%) off	Amount Pay (pay 80%)	Expression	New Price	Sales Tax (8%)	Overall Cost	Expression
100	20	80	$100 - 100(0.20) = 100(0.80)$				
50	10	40	$50 - 50(0.20) = 50(0.80)$				
28	5.60	22.40	$28 - 28(0.20) = 28(0.80)$				
14.50	2.90	11.60	$14.50 - 14.50(0.20)$ or $14.50(0.80)$				
x	$0.20x$	$x - 0.20x$	$x - 0.20x$ or $0.80x$				

Lesson 19:   Writing, Evaluating, and Finding Equivalent Expressions with Rational Numbers

© 2019 Great Minds®. eureka-math.org

Lesson Summary

- Two expressions are equivalent if they yield the same number for every substitution of numbers for the letters in each expression.

- The expression that allows us to find the cost of an item after the discount has been taken and the sales tax has been added is written by representing the discount price added to the discount price multiplied by the sales tax rate.

Name _____   Date _____

1. Write three equivalent expressions that can be used to find the final price of an item costing g dollars that is on sale for 15% off and charged 7% sales tax.

2. Using all of the expressions, determine the final price for an item that costs $75. If necessary, round to the nearest penny.

3. If each expression yields the same final sale price, is there anything to be gained by using one over the other?

4. Describe the benefits, special characteristics, and properties of each expression.

Solve the following problems. If necessary, round to the nearest penny.

1. Viviana is having her birthday party at the movie theater. Eight total people attended the party, and Viviana's parents bought each person a ticket. Viviana's parents also bought all of the drinks and snacks; four people chose a soda, two people chose a slushie, and six people chose a large popcorn.

   a. Write an expression that can be used to figure out the cost of the birthday party. Include the definitions for the variables Viviana's mom may have used.

   Let t represent the cost of a movie ticket, s represent the cost of a soda, d represent the cost of a slushie, and p represent the cost of a large popcorn.

   $8t + 4s + 2d + 6p$

   > I know that 8 movie tickets were bought, so I multiply 8 by the price of a movie ticket. I follow this same process for the other terms in the expression.

   b. Viviana's dad wrote down $2(4t + 2s + 3p)$ to determine the total cost of the party. Was he correct? Explain why or why not?

   The expression Viviana's dad wrote is not correct. Although it is possible to apply the distributive property and factor out a 2, he did not complete the process correctly because he dropped a term. There is no longer a term to represent the cost of the slushies. Instead the expression should be $2(4t + 2s + d + 3p)$.

   c. What is the cost of the party if a movie ticket costs $9.25, a soda costs $3.25, a slushie costs $2.50, and a large popcorn costs $5.50?

   > I can substitute the known values into the expression from part (a) or the corrected expression in part (b).

   $8t + 4s + 2d + 6p$

   $8(9.25) + 4(3.25) + 2(2.50) + 6(5.50)$

   $74 + 13 + 5 + 33$

   125

   The cost of the birthday party is $125.

2. Benson started a resume business. He helps clients create great resumes when they are looking for new jobs. Benson charges each client $50 to create a resume but paid $110 for new computer software to help create fancy resumes.

    a. Write an expression to determine Benson's take-home pay after expenses.

       Let r represent the number of resumes Benson creates.

       $50r - 110$ Benson's only expense is the computer software program.

    b. If Benson help 8 clients create resumes, what was his take-home pay after expenses?

       $50r - 110$

       $50(8) - 110$

       290

       Benson's take-home pay would be* $290 *after paying his expenses.

3. Mr. and Mrs. Slater bought some new pillows and used a 10% off coupon that allowed them to save some money. Mr. Slater added the 8% sales tax to the original cost first, and then deducted the coupon savings. Mrs. Slater thought they would save more money if the final cost of the pillows was calculated by deducting the savings from the coupon and then adding the 8% sales tax to the reduced cost.

    a. Write an expression to represent each person's scenario if the original price of the pillows was p dollars.

The sales tax increases the cost of the pillows by 8% of the original cost of the pillows. The coupon decreases the cost of the pillows by 10% of the price of the pillows, including sales tax.

Mrs. Slater

$(p - 0.10p) + 0.08(p - 0.10p)$

$1.08(p - 0.10p)$

$1.08(0.90p)$

The coupon decreases the cost of the pillows by 10% of the original cost of the pillows. The sales tax increases the cost of the pillows by 8% of the reduced price of the pillows.

Mr. Slater

$(p + 0.08p) - 0.10(p + 0.08p)$

$0.90(p + 0.08p)$

$0.90(1.08p)$

    b. Explain how both of the expressions are equivalent.

       The three factors in the expression are the same, which means the two expressions are equivalent because multiplication is commutative. Both expressions will evaluate to the same value when any number is substituted in for p.

    Lesson 19:   Writing, Evaluating, and Finding Equivalent Expressions
                 with Rational Numbers

EUREKA MATH

© 2019 Great Minds®. eureka-math.org

Solve the following problems. If necessary, round to the nearest penny.

1. A family of 12 went to the local Italian restaurant for dinner. Every family member ordered a drink and meal, 3 ordered an appetizer, and 6 people ordered cake for dessert.

    a. Write an expression that can be used to figure out the cost of the bill. Include the definitions for the variables the server used.

    b. The waitress wrote on her ordering pad the following expression: $3(4d + 4m + a + 2c)$. Was she correct? Explain why or why not.

    c. What is the cost of the bill if a drink costs $3, a meal costs $20, an appetizer costs $5.50, and a slice of cake costs $3.75?

    d. Suppose the family had a 10% discount coupon for the entire check and then left an 18% tip. What is the total?

2. Sally designs web pages for customers. She charges $135.50 per web page; however, she must pay a monthly rental fee of $650 for her office. Write an expression to determine her take-home pay after expenses. If Sally designed 5 web pages last month, what was her take-home pay after expenses?

3. While shopping, Megan and her friend Rylie find a pair of boots on sale for 25% off the original price. Megan calculates the final cost of the boots by first deducting the 25% and then adding the 6% sales tax. Rylie thinks Megan will pay less if she pays the 6% sales tax first and then takes the 25% discount.

    a. Write an expression to represent each girl's scenario if the original price of the boots was x dollars.

    b. Evaluate each expression if the boots originally cost $200.

    c. Who was right? Explain how you know.

    d. Explain how both girls' expressions are equivalent.

Mathematical Modeling Exercise: College Investments

Justin and Adrienne deposited $20,000 into an investment account for 5 years. They hoped the money invested and the money made on their investment would amount to at least $30,000 to help pay for their daughter's college tuition and expenses. The account they chose has several benefits and fees associated with it. Every 6 months, a summary statement is sent to Justin and Adrienne. The statement includes the amount of money either gained or lost. Below are semiannual (twice a year) statements for a period of 5 years. In addition to the statements, the following information is needed to complete the task:

- For every statement, there is an administrative fee of $15 to cover costs such as secretarial work, office supplies, and postage.
- If there is a withdrawal made, a broker's fee is deducted from the account. The amount of the broker's fee is 2% of the transaction amount.

TASK: Using the above information, semiannual statements, register, and beginning balance, do the following:

1. Record the beginning balance and all transactions from the account statements into the register.
2. Determine the annual gain or loss as well as the overall 5-year gain or loss.
3. Determine if there is enough money in the account after 5 years to cover $30,000 of college expenses for Justin and Adrienne's daughter. Write a summary to defend your answer. Be sure to indicate how much money is in excess, or the shortage that exists.
4. Answer the related questions that follow.

College Investment Fund Semi-Annual Statement

January 1, 2008 – June 30, 2008

Investment Gain/(Loss) :   700.00

College Investment Fund Semi-Annual Statement

July 1, 2008 – December 31, 2008

Investment Gain/(Loss):   754.38

College Investment Fund Semi-Annual Statement

January 1, 2009 – June 30, 2009

Investment Gain/(Loss):   (49.88)

College Investment Fund Semi-Annual Statement

July 1, 2009 – December 31, 2009

Withdrawal: 500.00
Investment Gain/(Loss):   (17.41)

College Investment Fund Semi-Annual Statement

January 1, 2010 – June 30, 2010

Investment Gain/(Loss):   676.93

College Investment Fund Semi-Annual Statement

July 1, 2010 – December 31, 2010

Investment Gain/(Loss):   759.45

College Investment Fund Semi-Annual Statement

January 1, 2011 – June 30, 2011

Deposit: 1,500.00
Investment Gain/(Loss):   880.09

College Investment Fund Semi-Annual Statement

July 1, 2011 – December 31, 2011

Investment Gain/(Loss) :   922.99

College Investment Fund Semi-Annual Statement

January 1, 2012 – June 30, 2012

Deposit: 800.00
Investment Gain/(Loss):   942.33

College Investment Fund Semi-Annual Statement

July 1, 2012 – December 31, 2012

Investment Gain/(Loss):   909.71

Lesson 20:   Investments—Performing Operations with
                 Rational Numbers

EUREKA MATH

5. Register

DATE	DESCRIPTION OF TRANSACTION	WITHDRAWAL	DEPOSIT	BALANCE	EXPRESSION
	Beginning Balance	---	---	$20,000.00	$20,000.00
Jan. –June: 2008					
July – Dec.: 2008					
Jan. – June: 2009					
July – Dec.: 2009					
Jan. – June: 2010					
July – Dec.: 2010					
Jan. – June: 2011					
July – Dec.: 2011					
Jan. – June: 2012					
July – Dec.: 2012					

6. Annual Gain/Loss Summary

Year	Total Gain/(Loss)	Numerical Expression
2008		
2009		
2010		
2011		
2012		
5-Year Gain/Loss		

7. Summary

Investments—Performing Operations with Rational Numbers

EUREKA
MATH

8. Related Questions

  a. For the first half of 2009, there was a $700 gain on the initial investment of $20,000. Represent the gain as a percentage of the initial investment.

  b. Based on the gains and losses on their investment during this 5-year period, over what period of time was their investment not doing well? How do you know? What factors might contribute to this?

  c. In math class, Jaheim and Frank were working on finding the total amount of the investment after 5 years. As a final step, Jaheim subtracted $150 for administrative fees from the balance he arrived at after adding in all the deposits and subtracting out the one withdrawal and broker's fee. For every semiannual statement, Frank subtracted $15 from the account balance for the administrative fee. Both boys arrived at the same ending 5-year balance. How is this possible? Explain.

  d. Based on the past statements for their investment account, predict what activity you might expect to see on Adrienne and Justin's January–June 2013 account statement. Then record it in the register to arrive at the balance as of June 30, 2013.

  e. Using the answer from part (d), if their daughter's college bill is due in September 2013 of, how much money do you estimate will be in their investment account at the end of August 2013 before the college bill is paid? Support your answer.

Exercise

Below is a transaction log of a business entertainment account. The transactions are completed and the ending balance in the account is $525.55. Determine the beginning balance.

DATE	DESCRIPTION OF TRANSACTION	PAYMENT	DEPOSIT	BALANCE
	Beginning Balance	---	---	
12/1/10	Bargain Electronic (i-Pod)	199.99		
12/5/10	Lenny's Drive-Up (Gift Certificate)	75.00		
12/7/10	Check from Customer: Reynolds		200.00	
12/15/10	Pasta House (Dinner)	285.00		
12/20/10	Refund from Clear's Play House		150.00	
12/22/10	Gaffney's Tree Nursery	65.48		525.55

Lesson 20:   Investments—Performing Operations with Rational Numbers

© 2019 Great Minds®. eureka-math.org

EUREKA MATH

Lesson Summary

- Calculations with rational numbers are used when recording investment transactions.

- Deposits are added to an account balance; money is deposited into the account.

- Gains are added to an account balance; they are positive returns on the investment.

- Withdrawals are subtracted from an account balance; money is taken out of the account.

- Losses are subtracted from an account balance; they are negative returns on the investment.

- Fees are subtracted from an account balance; the bank or financial company is charging you for a service.

Name _____   Date _____

1. Using the incomplete register below, work forward and backward to determine the beginning and ending balances after the series of transactions listed.

DATE	DESCRIPTION OF TRANSACTION	PAYMENT	DEPOSIT	BALANCE
	Beginning Balance	---	---	
1/31/12	Paycheck		350.55	
2/1/2012	Gillian's Chocolate Factory (Candy)	32.40		685.26
2/4/12	Main Street Jeweler's	425.30		
2/14/12	Saratoga Steakhouse	125.31		

2. Write an expression to represent the balance after the paycheck was deposited on 1/31/12. Let x represent the beginning balance.

3. Write a numerical expression to represent the balance after the transaction for Main Street Jeweler's was made.

A high school basketball team is hosting a family night at one of their games to raise money for new uniforms.

a. The game will be played on January 12, and the cost of admission is $4. Write an expression to represent the total amount of money collected for admission. Evaluate the expression if 476 people attend the basketball game.

Let p represent the number of people who attended the basketball game.

4p

4(476)

This amount represents the deposit in the transaction log in part (c).

1,904

If 476 people attend the basketball game, $1,904 would be collected from admission.

b. The following expenses were necessary for the basketball game, and checks were written to pay each company.

- Referees for the game: *High School Referees Inc.* costs $104 and is paid for on January 8.
- T-Shirts from *T-Shirt World* for the first 50 fans: Cost of the t-shirts was $5.75 each plus 6% sales tax, and the t-shirts were bought on January 5.

Write a numerical expression to determine the cost of the t-shirts.

This value represents one of the payments in the transaction log in part (c).

50(5.75 + 5.75(0.06))

50(5.75 + 0.345)

50(6.095)

I add the sales tax to the original cost of each t-shirt to determine the total cost of each t-shirt and then multiply by the number of shirts needed.

304.75

The cost for the t-shirts is $304.75.

EUREKA
MATH®

Lesson 20:   Investments—Performing Operations with
             Rational Numbers

255

© 2019 Great Minds®. eureka-math.org

c. Complete the transaction log below based on the information presented in parts (a) and (b).

> I subtract the payments from the balance, and I add the deposit to the balance.

Date	Description of Transaction	Payment ($)	Deposit ($)	Balance ($)
	Beginning Balance			876.54
January 5	T-Shirt World	304.75		571.79
January 8	High School Referees Inc.	104.00		467.79
January 12	Game Admission		1,904.00	2,371.79

Analyze the results.

d. Write an expression to represent the profit earned from the family night. Use the expression to determine the profit if 476 people attend the basketball game.

Let p represent the number of people who attended the basketball game.

$4p - 104 - 304.75$

$4p - 408.75$

> The profit is the amount of money collected in admissions minus all the expenses (cost of the t-shirts and the referees).

$4(476) - 408.75$

$1,495.25$

> The profit matches the transaction log if I calculate the difference between the ending balance and the beginning balance.
>
> $2,371.79 - 876.54 = 1,495.25$

The profit if 476 people attend the basketball game is $1,495.25.

Lesson 20: Investments—Performing Operations with Rational Numbers

© 2019 Great Minds®. eureka-math.org

EUREKA MATH

1. You are planning a fundraiser for your student council. The fundraiser is a Glow in the Dark Dance. Solve each entry below, and complete the transaction log to determine the ending balance in the student account.

   a. The cost of admission to the dance is $7 per person, and all tickets were sold on November 1. Write an expression to represent the total amount of money collected for admission. Evaluate the expression if 250 people attended the dance.

   b. The following expenses were necessary for the dance, and checks were written to each company.

      ▪ DJ for the dance—*Music Madness DJ* costs $200 and paid for on November 3.
      ▪ Glow sticks from *Glow World, Inc.* for the first 100 entrants. Cost of glow sticks was $0.75 each plus 8% sales tax and bought on November 4.

      Complete the transaction log below based on this information

DATE	DESCRIPTION OF TRANSACTION	PAYMENT	DEPOSIT	BALANCE
	Beginning Balance	---	---	1,243.56

   c. Write a numerical expression to determine the cost of the glow sticks.

   Analyze the results.

   d. Write an algebraic expression to represent the profit earned from the fundraiser. (Profit is the amount of money collected in admissions minus all expenses.)

   e. Evaluate the expression to determine the profit if 250 people attended the dance. Use the variable p to represent the number of people attending the dance (from part (a)).

   f. Using the transaction log above, what was the amount of the profit earned?

2. The register below shows a series of transactions made to an investment account. Vinnie and Anthony both completed the register in hopes of finding the beginning balance. As you can see, they do not get the same answer. Who was correct? What mistake did the other person make? What was the monthly gain or loss?

Original Register

DATE	DESCRIPTION OF TRANSACTION	PAYMENT	DEPOSIT	BALANCE
	Beginning Balance	---	---	
3/1/11	Broker's Fee	250.00		
3/10/11	Loan Withdrawal	895.22		
3/15/11	Refund – Misc. Fee		50.00	
3/31/11	Investment Results		2,012.22	18,917.00

Vinnie's Work

DATE	DESCRIPTION OF TRANSACTION	PAYMENT	DEPOSIT	BALANCE
	Beginning Balance	---	---	18,000.00
3/1/11	Broker's Fee	250.00		17,750.00
3/10/11	Loan Withdrawal	895.22		16,854.78
3/15/11	Refund – Misc. Fee		50.00	16,904.78
3/31/11	Investment Results		2,012.22	18,917.00

Anthony's Work

DATE	DESCRIPTION OF TRANSACTION	PAYMENT	DEPOSIT	BALANCE
	Beginning Balance	---	---	19,834.00
3/1/11	Broker's Fee	250.00		20,084.00
3/10/11	Loan Withdrawal	895.22		20,979.22
3/15/11	Refund – Misc. Fee		50.00	20,929.22
3/31/11	Investment Results		2,012.22	18,917.00

EUREKA
MATH

Exploratory Challenge: Integer Game Revisited

Let's investigate what happens if a card is added or removed from a hand of integers.

My cards:

My score:

Event 1

My new score:

Conclusion:

Event 2

My new score:

Conclusion:

Event 3

My new score:

Expression:

Conclusion:

Lesson 21:   If–Then Moves with Integer Number Cards

EUREKA MATH

Event 4

Expression:

Conclusion:

Exercises

1. The table below shows two hands from the Integer Game and a series of changes that occurred to each hand. Part of the table is completed for you. Complete the remaining part of the table; then summarize the results.

	Hand 1	Result	Hand 2	Result
Original	$1 + (-4) + 2$		$0 + 5 + (-6)$	
Add 4	$1 + (-4) + 2 + 4$			
Subtract 1	$1 + (-4) + 2 + 4 - 1$			
Multiply by 3				
Divide by 2				

2. Complete the table below using the multiplication property of equality.

	Original expression and result	Equivalent expression and result
	$3 + (-5) =$	
Multiply both expressions by -3		
Write a conclusion using if–then		

Lesson Summary

- If a number sentence is true, and the same number is added to both sides of the equation, then the resulting number sentence is true. *(addition property of equality)*

- If a number sentence is true, and the same number is subtracted from both sides of the equation, then the resulting number sentence is true. *(subtraction property of equality)*

- If a number sentence is true, and both sides of the equation are multiplied by the same number, then the resulting number sentence is true. *(multiplication property of equality)*

- If a number sentence is true, and both sides of the equation are divided by the same nonzero number, then the resulting number sentence is true. *(division property of equality)*

Name _____   Date _____

Compare the two expressions: Expression 1: $6 + 7 + -5$

Expression 2: $-5 + 10 + 3$

1. Are the two expressions equivalent? How do you know?

2. Subtract -5 from each expression. Write the new numerical expression, and write a conclusion as an if–then statement.

3. Add 4 to each expression. Write the new numerical expression, and write a conclusion as an if–then statement.

4. Divide each expression by -2. Write the new numerical expression, and write a conclusion as an if–then statement.

1. Evaluate each expression.

   a. $2 + (-5) + (-8)$ **−11**
   b. $125 \div (-5) \times 4$ **−100**
   c. $-10 + 45 \times (-2)$ **−100**

   > I must follow the order of operations when evaluating expressions.

2. Which expressions from Problem 1 are equivalent?

   Expressions (b) and (c) are equivalent expressions because they both evaluate to the same value.

3. If the two equivalent expressions from Problem 1 are multiplied by 6, write an if–then statement using the properties of equality.

   If $125 \div (-5) \times 4 = -10 + 45 \times (-2)$, ***then*** $6(125 \div (-5) \times 4) = 6(-10 + 45 \times (-2))$.

   > If I make the same changes to equivalent expressions, then the resulting expressions will still be equivalent.

4. Simplify the expression.

   $3 + (-14) \times 2 \div (-7) - 21$

   $3 + (-28) \div (-7) - 21$

   $3 + 4 - 21$

   -14

   Using the expression, write an equation.

   $3 + (-14) \times 2 \div (-7) - 21 = -14$

   Rewrite the expression if 4 is subtracted from both expressions.

   $3 + (-14) \times 2 \div (-7) - 21 - 4 = -14 - 4$

   Write an if–then statement using the properties of equality.

   If $3 + (-14) \times 2 \div (-7) - 21 = -14$, ***then***
   $3 + (-14) \times 2 \div (-7) - 21 - 4 = -14 - 4$.

1. Evaluate the following numerical expressions.

    a. $2 + (-3) + 7$

    b. $-4 - 1$

    c. $-\dfrac{5}{2} \times 2$

    d. $-10 \div 2 + 3$

    e. $\left(\dfrac{1}{2}\right)(8) + 2$

    f. $3 + (-4) - 1$

2. Which expressions from Exercise 1 are equal?

3. If two of the equivalent expressions from Exercise 1 are divided by 3, write an if–then statement using the properties of equality.

4. Write an if–then statement if -3 is multiplied to the following equation: $-1 - 3 = -4$.

5. Simplify the expression. $5 + 6 - 5 + 4 + 7 - 3 + 6 - 3$

    Using the expression, write an equation.

    Rewrite the equation if 5 is added to both expressions.

    Write an if–then statement using the properties of equality.

In this lesson, you will transition from solving equations using tape diagrams to solving equations algebraically by *making zero* (using the additive inverse) and *making one* (using the multiplicative inverse). Justify your work by identifying which algebraic property you used for each step in solving the problems. Explain your work by writing out how you solved the equations step by step and relate each step to those used with a tape diagram.

Example 1: Yoshiro's New Puppy

Yoshiro has a new puppy. She decides to create an enclosure for her puppy in her backyard. The enclosure is in the shape of a hexagon (six-sided polygon) with one pair of opposite sides running the same distance along the length of two parallel flower beds. There are two boundaries at one end of the flower beds that are 10 ft. and 12 ft., respectively, and at the other end, the two boundaries are 15 ft. and 20 ft., respectively. If the perimeter of the enclosure is 137 ft., what is the length of each side that runs along the flower bed?

Example 2: Swim Practice

Jenny is on the local swim team for the summer and has swim practice four days per week. The schedule is the same each day. The team swims in the morning and then again for 2 hours in the evening. If she swims 12 hours per week, how long does she swim each morning?

EUREKA
MATH

Exercises

Solve each equation algebraically using if–then statements to justify each step.

1. $5x + 4 = 19$

2. $15x + 14 = 19$

3. Claire's mom found a very good price on a large computer monitor. She paid $325 for a monitor that was only $65 more than half the original price. What was the original price?

4. $2(x + 4) = 18$

5. Ben's family left for vacation after his dad came home from work on Friday. The entire trip was 600 mi. Dad was very tired after working a long day and decided to stop and spend the night in a hotel after 4 hours of driving. The next morning, Dad drove the remainder of the trip. If the average speed of the car was 60 miles per hour, what was the remaining time left to drive on the second part of the trip? Remember: Distance = rate multiplied by time.

EUREKA
MATH

Lesson Summary

We work backward to solve an algebraic equation. For example, to find the value of the variable in the equation $6x - 8 = 40$:

1. Use the addition property of equality to add the opposite of -8 to each side of the equation to arrive at
    $6x - 8 + 8 = 40 + 8$.

2. Use the additive inverse property to show that $-8 + 8 = 0$; thus, $6x + 0 = 48$.

3. Use the additive identity property to arrive at $6x = 48$.

4. Then use the multiplication property of equality to multiply both sides of the equation by $\frac{1}{6}$ to get:
    $$\left(\frac{1}{6}\right)6x = \left(\frac{1}{6}\right)48.$$

5. Then use the multiplicative inverse property to show that $\frac{1}{6}(6) = 1$; thus, $1x = 8$.

6. Use the multiplicative identity property to arrive at $x = 8$.

Name _____   Date _____

Susan and Bonnie are shopping for school clothes. Susan has $50 and a coupon for a $10 discount at a clothing store where each shirt costs $12.

Susan thinks that she can buy three shirts, but Bonnie says that Susan can buy five shirts. The equations they used to model the problem are listed below. Solve each equation algebraically, justify your steps, and determine who is correct and why.

<table>
<tr><td>Susan's Equation</td><td>Bonnie's Equation</td></tr>
<tr><td>$12n + 10 = 50$</td><td>$12n - 10 = 50$</td></tr>
</table>

For each equation below, explain the steps in determining the value of the variable. Then find the value of the variable, showing each step. Write if–then statements to justify each step in solving the equation.

1. $3(y - 2) = -15$

    Multiply both sides of the equation by $\frac{1}{3}$, and then add 2 to both sides of the equation, $y = -3$.

    If: $3(y - 2) = -15$

    Then: $\frac{1}{3}(3(y - 2)) = \frac{1}{3}(-15)$ ⟵ | I use the multiplication property of equality using the multiplicative inverse of 3.

    If: $1(y - 2) = -5$ ⟵ | I recognize the multiplicative identity.

    Then: $y - 2 = -5$

    If: $y - 2 = -5$ ⟵ | I use the addition property of equality by using the additive inverse of -2.

    Then: $y - 2 + 2 = -5 + 2$

    If: $y + 0 = -3$ ⟵ | I recognize the additive identity.

    Then: $y = -3$

2. $2 = \frac{3}{4}a + 8$

    Subtract 8 from both sides of the equation, and then multiply both sides of the equation by $\frac{4}{3}$, $a = -8$.

    If: $2 = \frac{3}{4}a + 8$ ⟵ | The variable is on the right side of the equation, so I need to look at the right side of the equation to determine how to solve.

    Then: $2 - 8 = \frac{3}{4}a + 8 - 8$

    If: $-6 = \frac{3}{4}a + 0$

    Then: $-6 = \frac{3}{4}a$

    If: $-6 = \frac{3}{4}a$

    Then: $\frac{4}{3}(-6) = \frac{4}{3}\left(\frac{3}{4}a\right)$ ⟵ | The multiplicative inverse of $\frac{3}{4}$ is $\frac{4}{3}$.

    If: $-8 = 1a$

    Then: $-8 = a$

For each problem below, explain the steps in finding the value of the variable. Then find the value of the variable, showing each step. Write if–then statements to justify each step in solving the equation.

1. $7(m + 5) = 21$

2. $-2v + 9 = 25$

3. $\frac{1}{3}y - 18 = 2$

4. $6 - 8p = 38$

5. $15 = 5k - 13$

Exercises

1. Youth Group Trip

    The youth group is going on a trip to an amusement park in another part of the state. The trip costs each group member $150, which includes $85 for the hotel and two one-day combination entrance and meal plan passes.

    a. Write an equation representing the cost of the trip. Let P be the cost of the park pass.

    b. Solve the equation algebraically to find the cost of the park pass. Then write the reason that justifies each step using if–then statements.

    c. Model the problem using a tape diagram to check your work.

Suppose you want to buy your favorite ice cream bar while at the amusement park and it costs $2.89. If you purchase the ice cream bar and 3 bottles of water, pay with a $10 bill, and receive no change, then how much did each bottle of water cost?

d. Write an equation to model this situation.

e. Solve the equation to determine the cost of one water bottle. Then write the reason that justifies each step using if–then statements.

f. Model the problem using a tape diagram to check your work.

EUREKA
MATH

2. Weekly Allowance

Charlotte receives a weekly allowance from her parents. She spent half of this week's allowance at the movies, but earned an additional $4 for performing extra chores. If she did not spend any additional money and finished the week with $12, what is Charlotte's weekly allowance?

a. Write an equation that can be used to find the original amount of Charlotte's weekly allowance. Let A be the value of Charlotte's original weekly allowance.

b. Solve the equation to find the original amount of allowance. Then write the reason that justifies each step using if–then statements.

c. Explain your answer in the context of this problem.

d. Charlotte's goal is to save $100 for her beach trip at the end of the summer. Use the amount of weekly allowance you found in part (c) to write an equation to determine the number of weeks that Charlotte must work to meet her goal. Let w represent the number of weeks.

e. In looking at your answer to part (d) and based on the story above, do you think it will take Charlotte that many weeks to meet her goal? Why or why not?

3. Travel Baseball Team

Allen is very excited about joining a travel baseball team for the fall season. He wants to determine how much money he should save to pay for the expenses related to this new team. Players are required to pay for uniforms, travel expenses, and meals.

a. If Allen buys 4 uniform shirts at one time, he gets a $10.00 discount so that the total cost of 4 shirts would be $44. Write an algebraic equation that represents the regular price of one shirt. Solve the equation. Write the reason that justifies each step using if–then statements.

Lesson 23:   Solving Equations Using Algebra

EUREKA
MATH

b. What is the cost of one shirt without the discount?

c. What is the cost of one shirt with the discount?

d. How much more do you pay per shirt if you buy them one at a time (rather than in bulk)?

Allen's team was also required to buy two pairs of uniform pants and two baseball caps, which total $68. A pair of pants costs $12 more than a baseball cap.

e. Write an equation that models this situation. Let *c* represent the cost of a baseball cap.

f. Solve the equation algebraically to find the cost of a baseball cap. Write the reason that justifies each step using if–then statements.

g. Model the problem using a tape diagram in order to check your work from part (f).

h. What is the cost of one cap?

i. What is the cost of one pair of pants?

Lesson 23:   Solving Equations Using Algebra

EUREKA MATH

Lesson Summary

Equations are useful to model and solve real-world problems. The steps taken to solve an algebraic equation are the same steps used in an arithmetic solution.

Name _____ Date _____

Andrew's math teacher entered the seventh-grade students in a math competition. There was an enrollment fee of $30 and also an $11 charge for each packet of 10 tests. The total cost was $151. How many tests were purchased?

Set up an equation to model this situation, solve it using if–then statements, and justify the reasons for each step in your solution.

1. Solve the equation algebraically using if–then statements to justify your steps.

$$5 = \frac{-5+d}{3}$$

If: $5 = \frac{-5+d}{3}$

Then: $3(5) = 3\left(\frac{-5+d}{3}\right)$

> Dividing by 3 is the same as multiplying by $\frac{1}{3}$.
> The multiplicative inverse of $\frac{1}{3}$ is 3.

If: $15 = 1(-5+d)$

Then: $15 = -5+d$

If: $15 = -5+d$

Then: $15 + 5 = -5 + 5 + d$

If: $20 = 0 + d$

Then: $20 = d$

For Problems 2–3, write an equation to represent each word problem. Solve the equation showing the steps, and then state the value of the variable in the context of the situation.

2. Julianne works two part-time jobs. She waters her neighbor's plants every day for 1 hour. Julianne also babysits every day. She continues the same schedule for 8 days and works a total of 32 hours. If Julianne babysits for the same number of hours every day, how many hours does she babysit each day?

Let b represent the number of hours Julianne babysits each day.

$8(b+1) = 32$

If: $8(b+1) = 32$

Then: $\frac{1}{8}\left(8(b+1)\right) = \frac{1}{8}(32)$

> I could have applied the distributive property and then used properties of equality to solve for the missing variable.

If: $1(b+1) = 4$

Then: $b + 1 = 4$

If: $b + 1 = 4$

> If more explanation on solving equations is needed, I can refer back to Lesson 22.

Then: $b + 1 - 1 = 4 - 1$

If: $b + 0 = 3$

Then: $b = 3$

Julianne babysits for 3 hours each day.

3. Vince is thinking of joining a new gym and would have to pay a $55 sign-up fee plus monthly payments of
    $35. If Vince can only afford to pay $265 for a gym membership, for how many months can he be a
    member?

    Let m represent the number of months Vince can afford to be a member of the gym.

    $55 + 35m = 265$

    If: $55 + 35m = 265$

    Then: $55 - 55 + 35m = 265 - 55$

    If: $0 + 35m = 210$

    Then: $35m = 210$

    If: $35m = 210$

    Then: $\frac{1}{35}(35m) = \frac{1}{35}(210)$

    If: $1m = 6$

    Then: $m = 6$

    Vince can be a member of the gym for 6 months.

    > I know the sign-up fee is a one-time
    > payment, and then Vince has to pay
    > $35 every month.

For Exercises 1–4, solve each equation algebraically using if–then statements to justify your steps.

1. $\frac{2}{3}x - 4 = 20$

2. $4 = \frac{-1+x}{2}$

3. $12(x + 9) = -108$

4. $5x + 14 = -7$

For Exercises 5–7, write an equation to represent each word problem. Solve the equation showing the steps and then state the value of the variable in the context of the situation.

5. A plumber has a very long piece of pipe that is used to run city water parallel to a major roadway. The pipe is cut into two sections. One section of pipe is 12 ft. shorter than the other. If $\frac{3}{4}$ of the length of the shorter pipe is 120 ft., how long is the longer piece of the pipe?

6. Bob's monthly phone bill is made up of a $10 fee plus $0.05 per minute. Bob's phone bill for July was $22. Write an equation to model the situation using m to represent the number of minutes. Solve the equation to determine the number of phone minutes Bob used in July.

7. Kym switched cell phone plans. She signed up for a new plan that will save her $3.50 per month compared to her old cell phone plan. The cost of the new phone plan for an entire year is $294. How much did Kym pay per month under her old phone plan?

Credits

Great Minds® has made every effort to obtain permission for the reprinting of all copyrighted material. If any owner of copyrighted material is not acknowledged herein, please contact Great Minds for proper acknowledgment in all future editions and reprints of this module.